Essential Techniques for Medical and Life Scientists: A Guide to Contemporary Methods and Current Applications with the Protocols

(Part 1)

Edited by

Yusuf Tutar

Faculty of Pharmacy, Division of Biochemistry
University of Health Sciences, Istanbul, Turkey(

**Essential Techniques for Medical and Life Scientists:
A Guide to Contemporary Methods and Current Applications
with the Protocols**

Part # 1

Editor: Yusuf Tutar

ISBN (Online): 978-1-68108-709-2

ISBN (Print): 978-1-68108-710-8

©2018, Bentham eBooks imprint.

Published by Bentham Science Publishers – Sharjah, UAE. All Rights Reserved.

First published in 2018.

General:

1. Any dispute or claim arising out of or in connection with this License Agreement or the Work (including non-contractual disputes or claims) will be governed by and construed in accordance with the laws of the U.A.E. as applied in the Emirate of Dubai. Each party agrees that the courts of the Emirate of Dubai shall have exclusive jurisdiction to settle any dispute or claim arising out of or in connection with this License Agreement or the Work (including non-contractual disputes or claims).
2. Your rights under this License Agreement will automatically terminate without notice and without the need for a court order if at any point you breach any terms of this License Agreement. In no event will any delay or failure by Bentham Science Publishers in enforcing your compliance with this License Agreement constitute a waiver of any of its rights.
3. You acknowledge that you have read this License Agreement, and agree to be bound by its terms and conditions. To the extent that any other terms and conditions presented on any website of Bentham Science Publishers conflict with, or are inconsistent with, the terms and conditions set out in this License Agreement, you acknowledge that the terms and conditions set out in this License Agreement shall prevail.

Bentham Science Publishers Ltd.
Executive Suite Y - 2
PO Box 7917, Saif Zone
Sharjah, U.A.E.
Email: subscriptions@benthamscience.org

BENTHAM SCIENCE

CONTENTS

FOREWORD

Evaluating hypotheses and making judgements on the basis of well supported evidence solve problems in science. Medical and biological data mainly consist of measurement of assay conditions to support the proposed hypothesis. A prediction must test variables and scientist must acquire data from experiments to elicit an explanation or description to the scientific problem. Experimental results of the assay provide direct or indirect evidence to the theory. Therefore, several techniques have been developed to help researchers facilitate experimental design and to elucidate molecular mechanism of biological systems. Some techniques alone may provide unique evidence while still other complement other techniques. For this reason, investigators must be informed about the techniques and know them in detail. It is hard to follow up each technique specially for early career scientists and its applications for experienced researchers. The first volume of this book provides a training platform and application modules of each technique and may help individual investigators to guide research practices.

Prof. Lütfi Tutar
Ahi Evran University,
Turkey

PREFACE

This book focuses on instrumental techniques and their applications in medicine and biological sciences. Chapter 1 discusses mass spectroscopy (MS) and this technique provides qualitative and quantitative measurements of biological samples. MS has several applications but have not been mainly employed in proteomics studies yet. However, recent developments in clinical applications of MS courage researchers to perform difficult assays precisely and rapidly.

Structural elucidation of macromolecules form the basis of molecular biophysics and Chapter 2 discusses X-ray crystallography, Nuclear Magnetic resonance (NMR), Small Angle X-ray scattering (SAXS), and Cryo-electron Microscopy. Without detailed structure of a macromolecule, it is hard to understand the macromolecule interaction in biochemistry. So far, these methods have been employed in biological systems to reveal macromolecule structures. Each method has a unique site as well as overlapping functions however, all of them are useful to structural biologists, pharmaceutical, and medical scientists.

Chapter 3 explains a unique technique; isothermal titration calorimetry (ITC). It is unique since the technique not only measures binding affinity but also it provides thermodynamic data. The thermodynamic data provides nature of interaction. ITC measures any interactions between different macromolecules, DNA, RNA, lipid, carbohydrate, protein (enzyme, antibody), and ligand of any type. The component of the assay may be more than two molecules and there is no molecular weight restriction for the assay. Since it measures heat differences as signal, opaque solutions and suspensions do not restrict measuring binding affinity. ITC also measures enzyme kinetics and the technique may also be coupled to spectroscopic techniques like fluorescence. ITC is an all in one instrument for scientist. The chapter discusses applications of ITC in different disciplines.

Chapter 4 describes a common instrument of life science laboratory; reverse transcription polymerase chain reaction. This powerful method has found applications in medical, diagnosis, and forensics. Methodology and applications of the technique are explained thoroughly in this chapter.

This book is designed not only for early career young scientists (graduate students or postdoctoral associates) but for scientists who are experts in a particular technique but want to use different applications for their experimental set up. Next volume of the book will provide chapters for different analytical techniques.

Dr. Tutar would like to acknowledge networking contribution by the COST Action CM1407 "Challenging organic syntheses inspired by nature - from natural products chemistry to drug discovery".

Prof. Yusuf Tutar
Faculty of Pharmacy, Division of Biochemistry
University of Health Sciences
Istanbul
Turkey

List of Contributors

Ana Luísa Carvalho	UCIBIO, REQUIMTE, Departamento de Química, Faculdade de Ciências e Tecnologia, Universidade Nova de Lisboa, 2829-516, Caparica, Portugal
Banu Bayram	Health Sciences Faculty, Nutrition and Dietetics Department UCIBIO, REQUIMTE, Departamento de Química, Faculdade de Ciências e Tecnologia, Universidade Nova de Lisboa, 2829-516, Caparica, Portugal
Elvan Yılmaz Akyüz	Health Sciences Faculty, Nutrition and Dietetics Department Health Sciences Faculty, Nutrition and Dietetics Department Plant Genetic Engineering Laboratory, Department of Biotechnology, Bharathiar University, Coimbatore, Tamil Nadu, India
Esen Tutar	Kahramanmaraş Sütçü Imam University, Science and Letters Faculty, Avsar Campus, 46060, Kahramanmaras, Turkey
Eurico J. Cabrita	Health Sciences Faculty, Nutrition and Dietetics Department UCIBIO, REQUIMTE, Departamento de Química, Faculdade de Ciências e Tecnologia, Universidade Nova de Lisboa, 2829-516, Caparica, Portugal
Filipa Marcelo	UCIBIO, REQUIMTE, Departamento de Química, Faculdade de Ciências e Tecnologia, Universidade Nova de Lisboa, 2829-516, Caparica, Portugal
Halime Hanım Pençe	Faculty of Medicine, Division of Biochemistry Plant Genetic Engineering Laboratory, Department of Biotechnology, Bharathiar University, Coimbatore, Tamil Nadu, India
Maria João Romão	UCIBIO, REQUIMTE, Departamento de Química, Faculdade de Ciências e Tecnologia, Universidade Nova de Lisboa, 2829-516, Caparica, Portugal
Özlem Aytekin	Health Sciences Faculty, Nutrition and Dietetics Department Health Sciences Faculty, Nutrition and Dietetics Department Plant Genetic Engineering Laboratory, Department of Biotechnology, Bharathiar University, Coimbatore, Tamil Nadu, India
Sathishkumar Ramalingam	Faculty of Medicine, Division of Biochemistry Plant Genetic Engineering Laboratory, Department of Biotechnology, Bharathiar University, Coimbatore, Tamil Nadu, India
Serap Pektas	Department of Chemistry, Recep Tayyip Erdogan University, Rize, Turkey
Teresa Santos-Silva	UCIBIO, REQUIMTE, Departamento de Química, Faculdade de Ciências e Tecnologia, Universidade Nova de Lisboa, 2829-516, Caparica, Portugal
Venkidasamy Baskar	Health Sciences Faculty, Nutrition and Dietetics Department Health Sciences Faculty, Nutrition and Dietetics Department Plant Genetic Engineering Laboratory, Department of Biotechnology, Bharathiar University, Coimbatore, Tamil Nadu, India
Yusuf Tutar	Faculty of Pharmacy, Division of Biochemistry University of Health Sciences 34668, Istanbul, Turkey

CHAPTER 1

Application of Mass Spectrometry in Proteomics

Serap Pektas[*]

Department of Chemistry, Recep Tayyip Erdogan University, Rize, Turkey

Abstract: Mass spectrometry (MS) is a powerful tool to study biological samples both qualitatively (structure) and quantitatively (molecular mass). In recent years with the improvement of soft ionization techniques and the development of new methods, its application in proteomics, microbiology and clinical laboratories has increased. Especially in proteomics laboratories, it is commonly used for determining protein amino acid sequence, identification of post translational modifications, determining protein-peptide/protein/DNA interactions, determining protein folding and unfolding rates *etc.* In microbiological laboratories, MS is mainly used to identify microorganisms such as bacteria and fungi. Even though its use in clinical laboratories still needs improvement of methods, it can be used for diagnosis of disease, identification of metabolic disorders, discovering new biomarkers and identifying drug toxicity. This chapter provides a general review of MS applications in proteomics.

Keywords: Chemical Cross Linking coupled by MS (XL-MS), Electrospray Ionization, Hydrogen Deuterium Exchange coupled by MS (HDX-MS), Ion Trap Mass Analyzer, Matrix Assisted Laser Desorption Ionization, Proteomics, Protein Sequencing, Quadrupole Mass Analyzer, Time of Flight Mass Analyzer.

INTRODUCTION

The sample of interest is introduced to the mass spectrometer using an inlet system then reaches into the ionization source, where they are ionized. The formed ions are then arrived to a mass analyzer. After the ions reach to the mass analyzer, they separated based on their mass/charge (m/z) ratios later transferred to a detector. Finally signals are recorded by a computer system. The signals displayed as a mass spectrum showing the relative abundance of signals according to their m/z ratios. A typical mass spectrometer consists of the following four main parts (Fig. **1**):

[*] **Corresponding author Serap Pektas:** Department of Chemistry, Recep Tayyip Erdogan University, Rize, Turkey; Tel: +90 464 223 6126 (1819); E-mail: serappektas@gmail.com

Yusuf Tutar (Ed.)

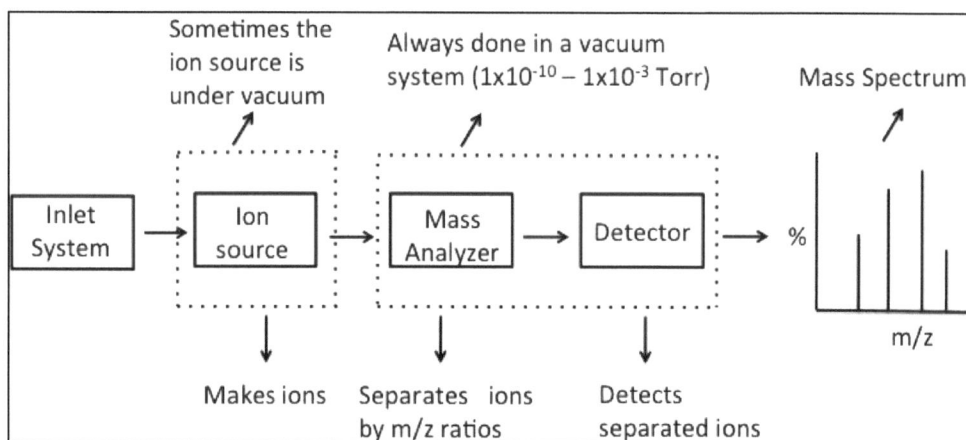

Fig. (1). Mass spectrometer parts.

1. Inlet system (LC, GC, Direct Probe)
2. Ion source (ESI, MALDI, FAB, CI, EI)
3. Mass analyzer (Quadrupole, Time of Flight (TOF), Ion Trap, Magnetic Sector)
4. Detector (Electron Multiplier, Micro Channel Plates MCPs)

1. INLET SYSTEMS

The function of an inlet system is to introduce a small amount of sample into the ion source. Sample can be injected to the mass spectrometer with different ways depend on the nature of the sample. One of them is *batch inlet* system, which involves the volatilization of sample externally, then gradually leakage of the volatilized sample into the ionization source. It is not suitable for the liquid samples that have boiling temperature of 500 °C [1]. Another inlet system is called *direct probe inlet*. *Direct probe inlet* is suitable for solid and nonvolatile liquids [2]. If the analyte is a mixture then sample can be introduced using one of the chromatographic techniques like liquid chromatography (LC), gas chromatography (GC) and capillary electrophoresis (CE) [2, 3]. The given name of a MS instrument may refer to inlet system that a mass spectrometer has, such as LC-MS, GC-MS and CE-MS.

2. IONIZATION TECHNIQUE

MS measures the masses of ions and therefore sample has to be ionized to be able to measure its mass. There are multiple ion sources and each of them has their own advantages and disadvantages depending on the analyte of interest. Ionization techniques used in MS are listed below:

1. Fast Atom Bombardment Ionization (FAB)
2. Electrospray Ionization (ESI)
3. Matrix Assisted Laser Desorption Ionization (MALDI)
4. Electron Ionization (Electron Impact Ionization) (EI)
5. Chemical Ionization (CI)
6. Native Ion Chemical Ionization

Depending on the chemical and physical properties of the sample of interest, different ionizations techniques can be used. One of the main factors for choosing the most suitable ionization source is the thermolability of the analyte. For non-thermolibale and volatile samples, electron ionization and/or chemical ionization techniques can be used. However, thermally liable and nonvolatile samples such as peptides, proteins, and other biological samples, softer ionization techniques are more suitable. The most common soft ionization sources used for the biological samples are ESI and MALDI [4, 5]. Therefore, these ionization techniques will be discussed in this chapter. The given name of a mass spectrometry technique is usually refers to the ionization method being used such as ESI-MS and MALDI-MS.

2.1. Electrospray Ionization (ESI)

Electrospray ionization (ESI) technique is one of the softest ionization techniques in MS. In the past few decades, it became an important technique in structural biology laboratories and clinical laboratories for qualitative and quantitative measurement of metabolites in a complex mixture of sample. Basic principle of ESI is outlined in Fig. (**2**), ESI technique can be divided into three main steps. The first step is the nebulization of the sample solution into electrically charged droplets. The second step is the release of ions from droplets and the final step is the transportation of ions to the mass analyzer [6 - 8].

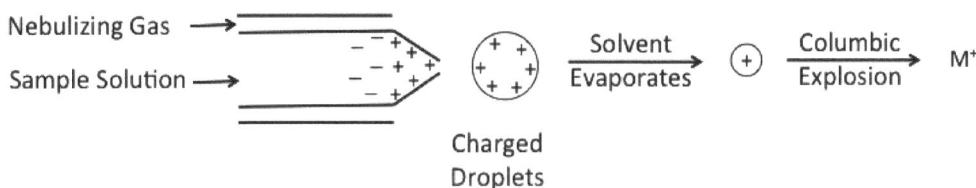

Fig. (2). Mechanism of Electrospray Ionization.

Sample solution passes through a capillary tube at a high voltage to generate ions. The applied potential between capillary tube and counter electrode is usually between 2.5-6 kV. After the initial formation of electrically charged droplets, they shrink in size by the evaporation of solvent, with the help of an ESI source temperature and/or nitrogen drying gas (nebulizer gas). The evaporation of solvent leads to a high charge density of droplets and coulomb repulsion force. When the electrostatic repulsion becomes stronger than the surface tension smaller electrically charged droplets are formed. Eventually ions at the surface of a droplet get ejected into the gaseous phase and then ions got accelerated into the mass analyzer [6]. However, for larger molecules like proteins, another model called charged residue mechanism is widely accepted. In this mechanism, due to the solvent evaporation and coulomb repulsion force, a very small charged droplet containing only a single analyte molecule is formed. After the desolvation of the charged droplet, its charge retains on the analyte molecule.

In ESI, technique charging is due to extra protons on analyte (or the loss of protons in negative mode) and compounds needs to have an acidic or basic charge to be ionized. One of the main advantages of ESI technique is that multiple charges can be generated which allow the determination of big molecules such as peptides and proteins.

Advantages:

• It is a very gentle ionization technique therefore it is suitable for biological molecules (proteins, peptides, *etc.*)
• Can analyze very large molecules
• Very sensitive and very efficient ionization technique
• Suitable to couple with LC systems.

Disadvantages:

• Low ionic strength solutions are necessary otherwise, the detector can be blocked and instrument may require maintenance (for protein and peptides only volatile buffers such as ammonium acetate can be used). But using low ionic strength solution may cause stability problems with some protein samples.
• Detergents are also not recommended because they suppress the ionization of analytes.
• It runs as continuous flow, which causes relatively more sample consumption compared to MALDI.

2.2. Matrix Assisted Laser Desorption Ionization (MALDI)

MALDI is a soft ionization technique used in MS. Thus, it can be used in the analysis of biomolecules such as peptides, proteins, DNA and sugars. In this technique, sample is co-crystalized with a UV absorbing substance called matrix (usually organic acids with conjugated pi system) [5, 9]. Ionization with MALDI requires three-step process. Initially, the sample is co crystalized with a suitable matrix and applied to a MALDI target (sample plate). Second, a laser pulse is used to irritate and trigger desorption of the analyte. Finally, the analyte molecules get ionized (Fig. **3**).

Laser Beam (hv)

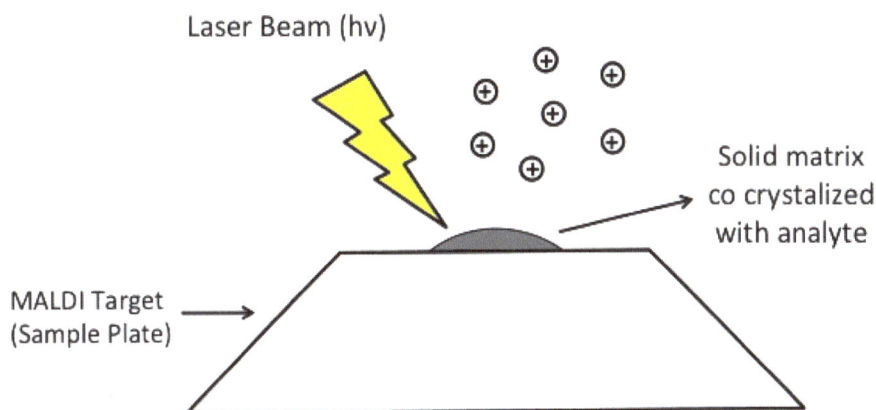

Solid matrix
co crystalized
with analyte

MALDI Target
(Sample Plate)

Fig. (3). Schematic diagram of a MALDI system.

Even though ionization mechanism by MADLI has not been well understood, matrix serves three main purposes. First, it absorbs laser energy and controllably transfers the energy. Second, it provides surface charging and finally keeps the analyte from aggregating. Table **1** shows the molecule of interests and appropriate matrix to use. As a laser source, nitrogen laser (337 nm) and frequency tripled and quadrupled Nd:YAG lasers (355nm and 266 nm, respectively) are commonly used [9]. Even though it is not very common, infrared lasers are also used in MALDI. One characteristic of MALDI is, that it commonly generates singly protonated molecules $(M + H)^+$.

Table 1. Molecule of interest and appropriate matrixes [9].

Molecule of Interest	Matrix	Abbreviation
Peptides/Proteins Mass < 10 000 Mass > 10 000	α-Cyano-4-hydroxycinnamic acid Sinapic acid 2,5-dihydroxybenzoic acid	CHCA SA HABA
Nucleotides Mass < 3.5 kDa Mass > 3.5 kDa	2,4,6-Trihydroxyacetophenone 3-Hydroxypicolinic acid Anthranilic acid Nicotinic acid Salicylamide	THAP HPA
Lipids	Dithranol	DIT

(Table 1) cont.....

Molecule of Interest	Matrix	Abbreviation
Carbohydrates	2,5-Dihydroxybenzoic acid α-Cyano-4-hydroxycinnamic acid 2,4,6-Trihydroxyacetophenone 3-Aminoquinoline	DHB CHCA THAP

2.2.1. MALDI Sample Preparation

Sample preparation can be divided into the following steps:

- Initially a solution is prepared using acetonitrile (ACN) (or methanol), water and trifluoroacetic acid (TFA) (or formic Acid, FA). The ratio of the solvent varies but 50:50:0.2 (ACN:Water:TFA) ratio is very common.
- After ACN:Water:TFA solution prepared, a saturating amount of MALDI matrix is added into the solution (it has to be saturated).
- Sample of interest is then added into the MALDI matrix solution and mixed well (with the help of a vortex).
- Then, 1-2 μL of matrix-protein mix is applied on MALDI target (sample plate). Finally, samples can be analyzed by MALDI-MS after samples are air-dried or blow-dried.

Advantages:

- A fast ionization technique
- Less sample loss due to the non-continuous flow
- More tolerant to salt concentration than ESI

Disadvantages:

- Non-continuous nature of the technique makes it difficult to couple with LC systems
- Not very convenient to perform protein MS/MS (for now)
- Not suitable for HDX-MS experiment due to the difficulty with temperature control (see section 5.1.4.1 of this chapter).

3. MASS ANALYZERS

A mass analyzer separates the ionized masses based on mass to charge (m/z) ratios, then a detector measures the signal. There are six types of mass analyzers that can be used in MS, each for different purposes. Among these mass analyzers, the following first four are commonly used in biological laboratories and they will be discussed in more details.

1. Quadrupole Mass Analyzer

2. Ion Trap Mass Analyzer
3. Orbitrap Mass Analyzer
4. Time of Flight Mass Analyzer
5. Magnetic Sector Mass Analyzer
6. Electrostatic Sector Mass Analyzer
7. Ion Cyclotron Resonance

3.1. Quadrupole Mass Analyzer

The quadrupole mass analyzer is one of the most commonly used mass analyzer in mass spectrometers. They are robust, economical, small size mass analyzers and suitable with different inlet systems.

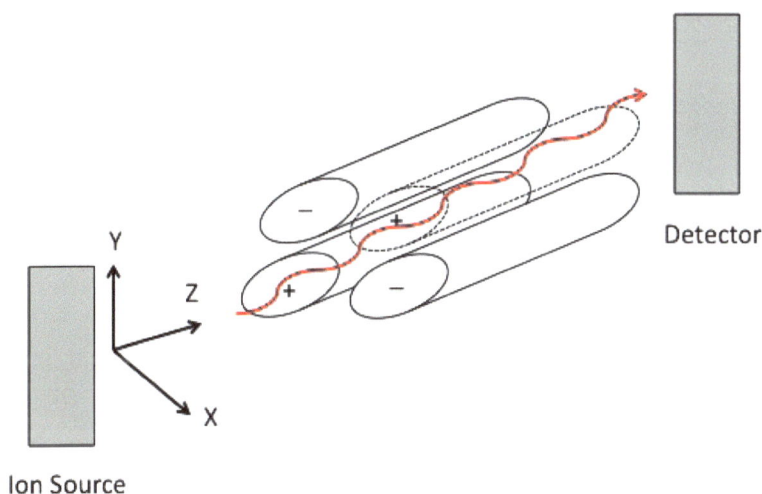

Fig. (4). Schematic diagram of a quadrupole mass analyzer.

Quadrupole system consists of 4 parallel metal rods placed in equal distance and each opposite rods connected with DC (direct current) and RF (radio frequency) voltages (Fig. **4**). The both voltages create an electrical field, which causes ions to travel with oscillatory motion in the Z direction. The oscillatory motion's amplitude is related with ions m/z ratio and allows selective transmission of ions [10 - 12]. By changing DC and RF voltages ions passing to detector can be controlled. If the ions are off the set m/z ratio limits, they hit the metal rods and cannot reach to the detector [10 - 12]. A mass spectrometer may have more than one quadrupole mass analyzer, for example, a tandem mass spectrometer may contain three quadrupole mass analyzers.

Tandem Mass Spectrometry MS/MS

A tandem mass spectrometry is applying two stages of mass analysis. It is also known as MS/MS or MS^2 (Fig. **5**). A tandem mass spectrometer has more than one mass analyzer, which allows the analysis of analyte at a structural level [2, 12 - 14].

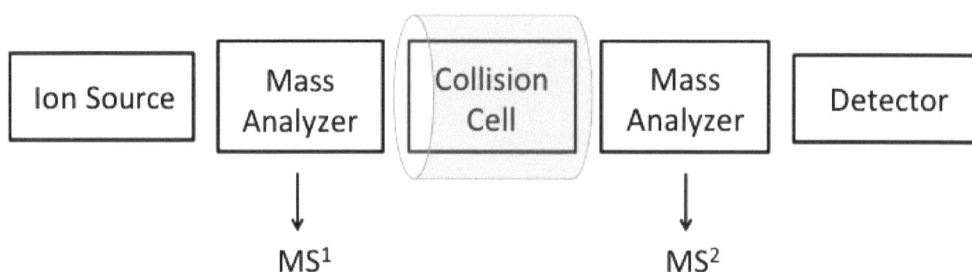

Ion Source	Mass Analyzer	Collision Cell	Mass Analyzer	Detector

MS^1 MS^2

Fig. (5). Tandem MS system.

The mass analyzers can be of same type as in QqQ (Triple Quadrupole) or could be of different type as in Q-TOF (Quadrupole Time of Flight) in later case, the mass spectrometer is called a hybrid one. Basis of tandem MS is after ions formed by an ion source, first mass analyzer isolates ions of a specific m/z value that corresponds to a single species (a peptide). The ion of interest is then mass-selected and this mass-selected ion is called precursor ion (or parent ion). Precursor ion gets fragmented in a collision tube with the help of a collision gas. By this process, some of the kinetic energy of the molecule converted into internal energy when the internal energy reaches to an access amount, it causes the breakage of bonds. This process is known as collision induced dissociation (CID) or collisional activation (CA). After fragmentation, the product ions (also called the daughter ions) transferred to another mass analyzer where they separated and analyzed based on their m/z ratios [15, 16]. CID is the major fragmentation method used for peptide sequencing in proteomics. However, it is not the method of choice for the analysis of the post translationally modified peptides and/or highly basic peptides. In this case, another method called electron transfer dissociation (ETD) is preferred. In ETD method, fragmentation is based on the transfer of an electron from an anion radical to a positively charged peptide. This induces the fragmentation of peptide backbone, which leads to dissociation of N-Cα [17, 18].

Tandem MS has wide range of applications from proteomics to medicine and some of them are discussed in applications section of this chapter.

3.2. Ion Trap Mass Analyzer (IT)

An ion trap mass analyzer consists of three hyperbolic electrodes: the central ring electrode and two adjacent end cap electrodes. Schematic diagram in Fig. (**6**) shows the assembly of an ion trap mass analyzer.

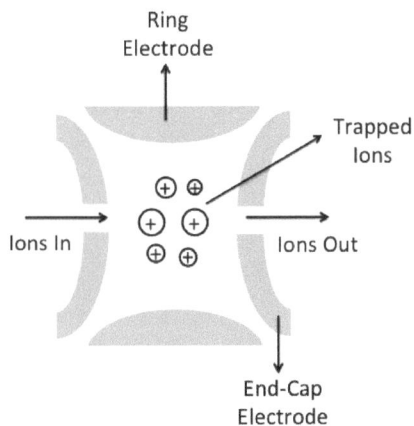

Fig. (6). Schematic diagram of an ion trap mass analyzer.

Positioning of the electrodes creates a cavity in which ions can get trapped (stored) and analyzed. The various voltages are applied to the electrodes to allow trapping and ejecting ions according to their m/z ratios [19 - 21]. ITs are small and cheap mass analyzers. One the benefits of ITs is, they can provide collision induced dissociation (CID) which is useful for tandem MS. CID is induced using an inert gas usually argon or nitrogen gas [7].

3.3. Orbitrap Mass Analyzer

The Orbitrap mass analyzer is relatively a new type of mass analyzer. As it name refers, Orbitrap is an ion trap but unlike traditional ion trap it does not contain magnets to trap the ions. Instead, ions are trapped in an electrostatic field [22]. The electrostatic field causes ions to move in spiral patterns. An Orbitrap mass analyzer consists of an internal electrode (central) and an external electrode (outer). The space between these two electrodes is called measurement chamber, which is under high vacuum condition (around 10^{-8} torr or lower) [22 - 24]. The main advantage of Orbitrap mass analyzer is it provides mass accuracy and high resolution.

3.4. Time of Flight (TOF) Mass Analyzer

Time of flight mass analyzer is one of the fastest mass analyzers. In this system, m/z ratio of an ion is determined by a time measurement. TOF uses kinetic energy and velocity of ions to separate them (Fig. **7**) [25 - 29].

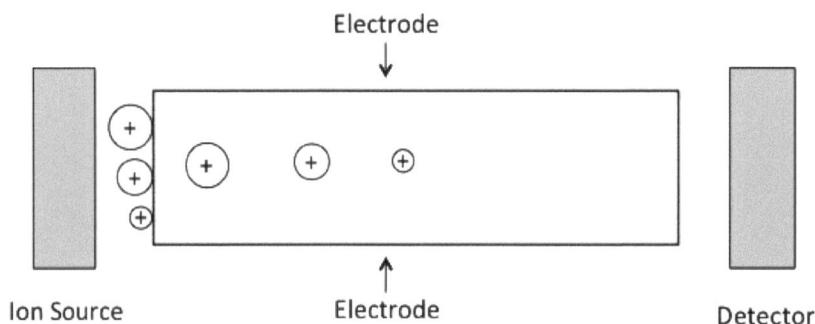

Fig. (7). Schematic diagram of a TOF mass analyzer.

The applied electric field causes all ions with the same charge to have same kinetic energies. Ions with the same kinetic energy and lesser m/z values have greater velocity and higher m/z values have lower velocity. As a result, ions get separated while travelling through the mass analyzer according to their m/z ratios and this is the basis of TOF mass analyzers. Time of flight is the time it takes ions move from an ion source to the detector. Ions of lesser m/z ratios arrive to the detector earlier than the ions of higher m/z ratios [25 - 29].

TOF is a sensitive mass analyzer but high masses require longer flight times, which cause resolution difficulties. To overcome this problem, usually a reflectron is added at the end of the flight tube (drift tube) of TOF analyzers [25, 30, 31]. The reflectron is made up of a series ring electrodes with a very high voltage. When ions reach to reflectron, high voltage causes ions to be reflected in opposite direction, back to the flight tube. The reflectron increases the resolution by broadening the range of flight times for m/z values.

4. DETECTORS

Electron Multiplier and Micro Channel Plates MCPs detectors are commonly used in MS instruments.

5. APPLICATIONS

Mass spectrometry is a very sensitive and reliable tool. It can provide qualitative and quantitative measurement for an analyte of interest. Over the past years with the improvement of soft ionization techniques, mass spectrometry has become a major tool in proteomic laboratories [32 - 35]. Recently, MS is also becoming a popular tool in microbiology laboratories for bacterial identification [36 - 39]. In addition to proteomics and microbiology laboratories, clinical laboratories also use MS for the diagnosis of disease, identification of metabolic disorders, discovering new biomarkers and identifying drug toxicity [40 - 44]. Some of these applications are described in details below.

5.1. Proteomics

Due to the role of proteins in many diseases, proteins are one of the main targets in drug discovery and extensive focus of research. Proteomics can be explained as an approach to study proteins in terms of cellular localization, structure, function and modification to understand biological processes. Today, MS is the most important tool in proteomics research and can address many questions including molecular weight determination, primary structure sequencing, post transitional modification, protein-protein/peptide/DNA/small molecule interactions, conformational changes, protein folding and unfolding rates. In this section, applications of MS to address these issues are summarized [32 - 35, 45 - 52].

5.1.1. Molecular Weight Determination

One application of MS in proteomics laboratories is determining the protein molecular weight. Amino acid composition determines the molecular weight and can be used to identify a protein. Traditional methods like immunoassays, sometimes may give false signals but MS provides more accurate results.

In order to get a good quality mass spectrum, sample preparation is important. Sample preparation is also depends on the instrumentation. If ESI-MS is the instrument choice then protein sample needs to be in a low ionic solvent such as ammonium acetate (10 mM), because high salt concentration may block the detector. Sample can be buffer exchanged by dialysis or desalting columns with a low ionic buffer but some proteins are not very stable at low ionic strength buffers and may aggregate. In this case a HPLC column can be used to remove the high salt. For this aim C4 and C8 columns can be used for intact proteins. Initially, protein sample ran through the HPLC column with an aqueous solvent (usually 0.1% formic acid (FA)) protein sample hold on the column surface and solvent washes out, after that, column connected to ESI-MS instrument. By running an organic solvent (usually 0.1% FA in acetonitrile (ACN)) through the column, protein samples come out and their masses measured. Fig. (**8**) shows the mass spectrum of purified HIF Prolyl Hydroxylase 2 (PHD2) [53].

If MALDI-MS is the instrument of choice then deciding the MALDI matrix is highly important. Table **1** shows the appropriate matrix based on molecular weight. If the mass of the protein is lower than · 10 000 Da, α-Cyano-4-hydroxycinnamic acid is used. If it is higher than 10 000 Da then Sinapic acid or 2,5-dihydroxybenzoic acid can be used. Since the MALDI is relatively salt tolerant instrument desalting step is not required. A matrix solution is prepared as described in section 2.2.1. After matrix solution prepared, protein sample is added into it and mixed with a vortex. Around 1-2 μL protein-matrix mixture spotted on MALDI target (MALDI plate). After that air or blow-dried sample is analyzed using MALDI-MS instrument.

Important Notice: Initially, calibration of the instrument is essential to obtain an accurate mass determination. One of the following standards can be used to calibrate MS instruments, Renin Substrate, Substance P, Cytochrome C, Apomyoglobin *etc*. After instrument calibrated sample can be introduced to the mass spectrometer depend on the method of choice.

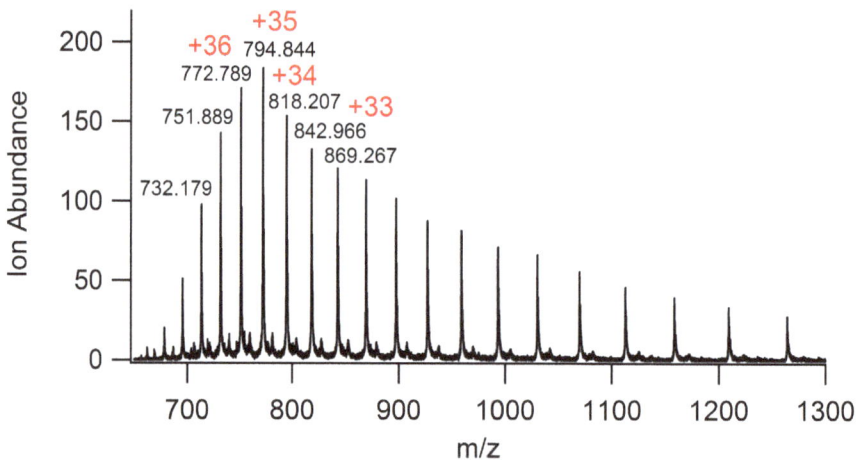

Fig. (8). The mass spectrum of the purified PHD2 enzyme (ESI-MS) [53].

When calculating molecules mass from mass spectrum, following assumptions are made:

- Charging is due to protons
- There is no such thing as partial protons
- Adjacent peaks in a mass spectrum differ by one charge.

The relation between two adjacent peaks in a mass spectrum can be given by following equations. In these equations "m" is the mass of the protein and "z_1" is the assigned charge of 818.207 peak.

$$\frac{m + z_1}{z_1} = 818.207 \tag{1}$$

$$\frac{m + (z_1 + 1)}{z_1 + 1} = 794.844 \tag{2}$$

By solving above equations (eq. 1 and eq.2), "z_1" is calculated as +34 and "m" calculated as 27785.04 Da. The charges of the other peaks can be calculated in the same way.

5.1.2. Protein Sequencing

One of the main uses of a mass spectrometer in proteomics laboratories is determining the primary structure of a protein (amino acid sequence). Primary structure is the key to identify a protein therefore, determination of a protein and/or discovering new proteins require sequence identification. For this purpose MS/MS instrumentation is required [48 - 51]. There are MS/MS instruments available with different ion sources, ESI-MS/MS and MALDI-MS/MS. Between these two systems, ESI-MS/MS is more convenient and provides better signals compared to MALDI-MS/MS system. In addition to different ionization source there are MS/MS instrument with different mass analyzers. In terms of mass analyzer choice, Q-TOF instruments have advantages over ion trap mass analyzers. The disadvantage of an ion trap instrument is that the lower mass end of the MS/MS spectrum is absent. On the other hand, Q-TOF instrument can provide a full scan.

In a MS/MS system, first mass analyzer takes a mass spectrum. Then a mass selected precursor ion is fragmented into the product ions. After the fragmentation a second mass spectrum is taken for product ions. The precursor ion and m/z differences between product ions give structural information. MS sequencing of a peptide is often called "*de novo* sequencing" and has done the following way.

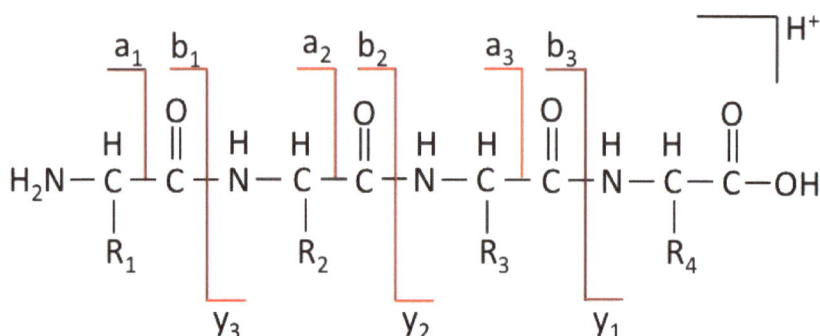

Fig. (9). Peptide fragmentation and ion labeling proposed by Biemann [48 - 51].

Peptide fragmentations occur randomly, meaning it does not start from N–terminus or C-terminus of a peptide. Fig. (**9**) shows the possible peptide fragmentations and labeling of each ion for sequencing [48 - 51, 54, 55]. Deciding which peak corresponds to which ion might look hard to decide but there are some rules to follow. Among these ions, "**b** ions" and "**y** ions" are commonly observed in collision-induced dissociation with a relatively good intensity. Frequency of "**a** ions" is usually low but it helps to identify "**b** ions". The "**a** ions" and "**b** ions" appear 28u away from each other due to the loss of carbonyl (C=O) and presence of two peaks 28u apart may indicate "**a** ions" and "**b** ions" [32, 48 - 51, 54, 55]. *De novo* sequencing is like solving a puzzle and gets better with practice. Figs. (**9** and **10**) are examples of how to perform "*de novo* sequencing". Given amino acid monoisotopic masses on Table **2** help to decide the amino acids.

Table 2. Amino acids and monoisotopic masses.

Amino Acid	Amino Acid Code	Monoisotopic Mass
Glycine	G	57.021464
Alanine	A	71.037114
Serine	S	87.032029
Proline	P	97.052764
Valine	V	99.068414
Threonine	T	101.04768
Cysteine	C	103.00919
Leucine	L	113.08406
Isoleucine	I	113.08406
Asparagine	N	114.04293

(Table 2) cont.....

Amino Acid	Amino Acid Code	Monoisotopic Mass
Aspartic acid	D	115.02694
Glutamine	Q	128.05858
Lysine	K	128.09496
Glutamic acid	E	129.04259
Methionine	M	131.04048
Histidine	H	137.05891
Phenylalanine	F	147.06841
Arginine	R	156.10111
Tyrosine	Y	163.06333
Tryptophan	W	186.07931

5.1.2.1. Sample Preparation for Protein Sequencing

The quality of the sample preparation significantly impacts the mass spectrum. In order to identify a protein amino acid sequence by MS/MS, a proper sample preparation is required. Since the proteome is very complex, there is not really a standard sample preparation method. It can be divided into the following basic steps:

- Proteins are usually present in a mixture other proteins. If the protein of interest is not pure, this causes major problems with MS analysis. One problem is that the abundant ions have the tendency to suppress the signal of less abundant ions. Another issue is that there will be an overwhelming number of species to analyze. In order to overcome this problem, proteins can be separated to some extend using an LC system, which will provide certain level of separation and will make the protein sample less complex.
- Protein samples are usually kept in high salt concentration to maintain structural stability. Therefore, an important consideration in sample preparation is, if ESI is the choice of ionization source, salt needs to be removed prior to mass ionization as explained in 5.1.1 section of this chapter. Dialysis or desalting columns can be used to achieve this.
- Whole protein sample is too large to perform MS/MS. To overcome this issue, protein of interest is digested into peptide fragments using a protease (trypsin or pepsin). Simply done by adding protease into protein sample then incubating it for overnight or couple of hours depending on the amount of protein and protease. Moreover, using an LC system minimizes the complexity of the sample to some extend (remember that LC can be coupled with ESI but not with MALDI).
- Cysteine residues may form disulfide bond, which create difficulty with

analysis. To prevent this, 2-Mercaptoethanol can be added into protein sample. This step is important, if the protein of interest contains Cysteine residue. This step can be done after digestion with a protease or before.

• After sample preparation, mass spectrum is taken by MS/MS instrument.

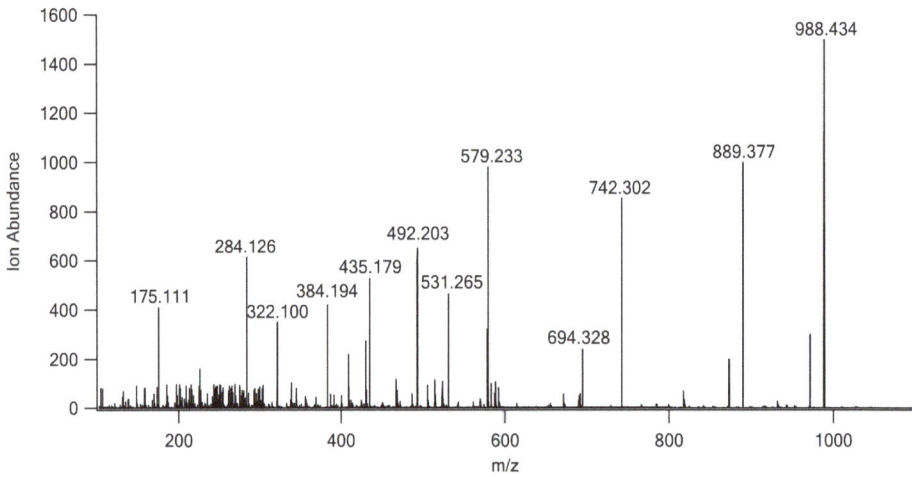

Fig. (10). A peptide MS/MS spectrum.

Fig. (**10**) is an example of a product ion spectrum, showing the product ions m/z ratios and relative abundance of these ions.

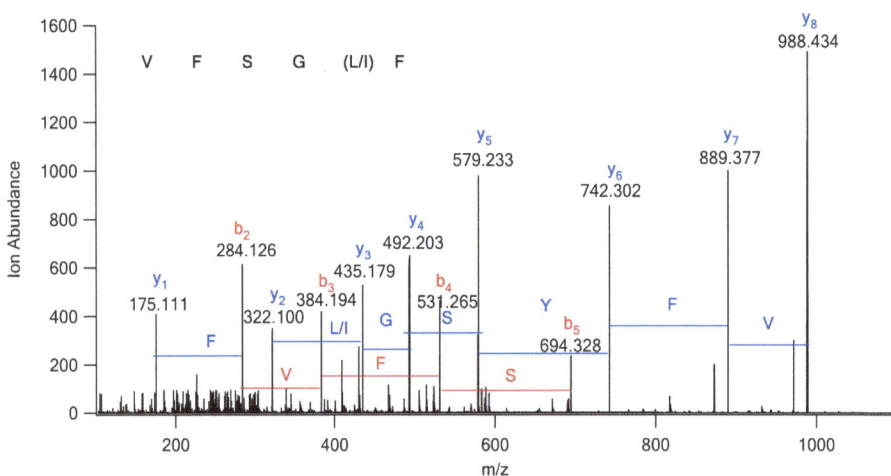

Fig. (11). A Peptide MS/MS spectrum with ion labeling [57-60].

In Fig. (**11**), product ions are labeled with a corresponding ion. This spectrum predominantly contains "**b** ions" and "**y** ions". The product ion peaks that extend from amino terminus are termed as "**b** ions" and the peaks extend from carboxyl terminus are termed as "**y** ions". Around 99 Da mass difference between "y_8" and "y_7" ions is the indication of Valine amino acid and also mass difference between "b_2" and "b_3" ions is indication of a Valine residue (Table **2** shows amino acids and their monoisotopic masses). As observed in Fig. (**11**) MS/MS spectrum is belong to the VFSG(L/I)F peptide fragment. One uncertainty in this peptide spectrum is Leucine or Isoleucine residue. They have the same exact monoisotopic masses, therefore it is a problem to decide which one is the MS/MS making it one drawback of the method.

5.1.3. Post-Translational Modifications

Proteins are subjected to a wide range of chemical modifications after translation. Some of these chemical modifications are phosphorylation, acetylation, hydroxylation, glycosylation, iodination, *etc.* and are also known as "post-translational modifications" (PTMs). PTMs are essential to initiate proteins catalytic function therefore, it is important to monitor. All these PTMs can be monitored using MS, by simply determining the gained mass upon chemical modification [32 - 35]. ESI-MS and MALDI-MS both can be used for this purpose. After protein of interest modified, sample preparation for MS can be performed as described in 5.1.1 section of this chapter. Initially, a mass spectrum is taken for intact protein to monitor if there is any covalent modification. A mass spectrum of intact protein can reveal whether there is a chemical modification or not but it can't show which amino acid modified. In order to determine the location of modification, MS/MS sequencing needs to be performed as described in section 5.1.2 of this chapter.

Fig. (**12**) shows the identification of hydroxylation using MALDI-TOF instrument. In this example CODD peptide is hydroxylated by PHD2 enzyme. The black spectrum was taken at the very beginning of the reaction therefore the amount of hydroxylated CODD was low. The gray mass spectrum is belongs to the same sample after 2 minutes exposure to the enzymatic modification. In addition to the identification of a PTM, by comparing these two spectrums the rate of this chemical modification can also be determined.

5.1.4. Protein-Protein, Protein-DNA and Protein-Small Molecule Interactions

Protein-protein, protein-peptide and protein-DNA interactions take place *in vivo* for a short period of time to either induce signaling or metabolic function. Therefore monitoring these interactions is important to understand biological processes. In proteomic laboratories a MS based method called hydrogen deuterium exchange coupled with MS (HDX-MS) is used to monitor these interactions [45, 46, 56]. Furthermore, proteins' interaction with biological molecules and/or small molecules can be monitored by HDX-MS method. Since proteins are the one of the main classes of drug targets monitoring protein-drug interactions are also essential. In addition to HDX-MS, protein-protein and protein-peptide interactions can be determined using another MS based method called chemical cross linker coupled with MS (XL-MS) [57 - 60]. These methods are described in details below.

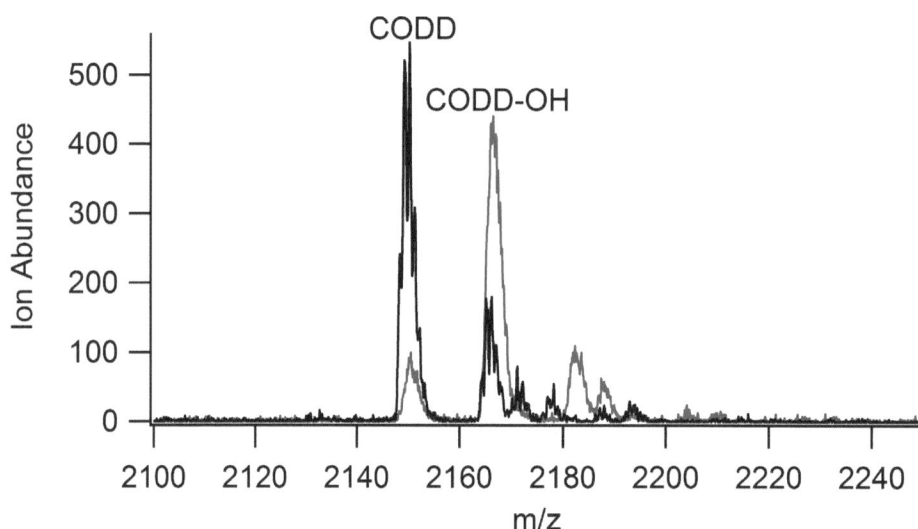

Fig. (12). Hydroxylation of CODD peptide monitored by MALDI-TOF [53].

5.1.4.1. Hydrogen Deuterium Exchange (HDX) Coupled with MS (HDX-MS)

The basis of hydrogen deuterium exchange (HDX) is solvent accessible hydrogens in the proteins can be exchanged with hydrogens from the solvent molecules. When D_2O is used as solvent, hydrogen will exchange with deuterium, and the protein will then gain mass which can be detected by mass spectrometry [40 - 43]. In proteins there are three different types of exchangeable hydrogens and each has a different exchange rate. The ones at the side chain of amino acids exchange is too fast that goes to back-exchange which makes it difficult to determine the exchange rate. Hydrogens of carbon-hydrogen bond are too slow to observe. On the other hand third type of hydrogen, the backbone amide hydrogens will exchange at a detectable rate and they are the interest for this technique [61, 65 - 67].

Exchange rate is directly related to solvent accessibility and hydrogen bonding, making amide HDX a sensitive reporter of both protein structure and flexibility.

Amide hydrogens that are easily accessible to the solvent will exchange quicker than the amide hydrogens in the core of the protein [41, 43]. Hydrogen bonding also affects the exchange rate, amide hydrogens which are a part of intramolecular hydrogen bonding (α-helix and β-sheet) will exchange very slowly [23, 34].

EX1 and EX2 Regimes

In HDX method, depending on the rates of protein refolding and intrinsic exchange, two distinct regimes are commonly observed, EX1 and EX2. Under native conditions, proteins usually follow EX2 regime in which the protein folding/unfolding rate is higher than the deuterium exchange rate. On the contrary, EX1 regime assumed to be observed in denaturing conditions [45, 61, 62, 67, 68]. EX1 kinetics regime takes place when a specific region of a protein is unfolded hydrogens in this area are exchange with D_2O before the protein refolds to its native state. In EX2 regime, protein/peptide mass continuously increase by time and appears as a single population. On the other hand, in EX1 regime, two mass populations are observed [45, 61, 66, 68].

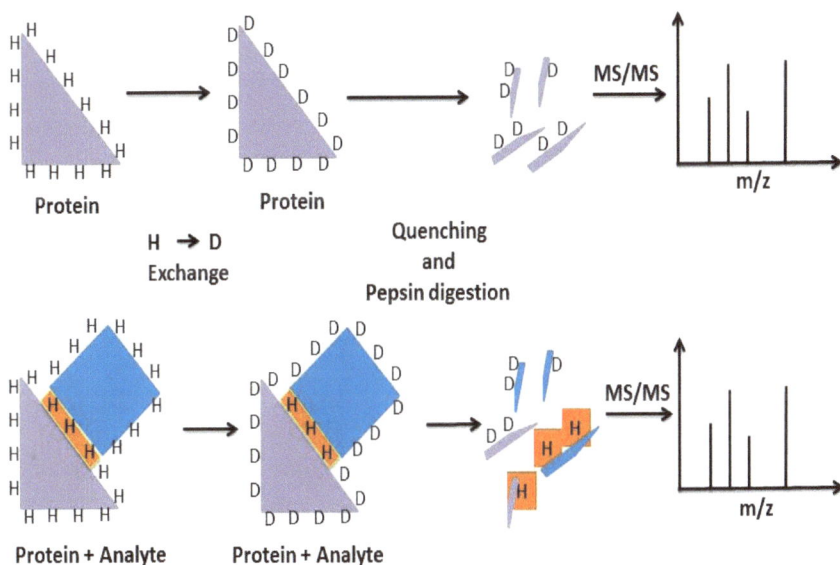

Fig. (13). Schematic diagram of HDX-MS method.

Fig. (**13**) summarizes the HDX-MS method. First, the protein of interest is exposed to D_2O, then the reaction is quenched at different time points. In order to identify which part of the protein is incorporated with deuterium, the protein is digested by pepsin. After proteolysis, the mass difference of the peptide fragments are monitored by MS/MS, this is called local HDX-MS. The exchange rate can be monitored by MS without proteolysis and is called global HDX-MS [61 - 63]. Usually global HDX-MS is performed before local HDX-MS to decide what to expect from local HDX-MS.

The exchange rate is highly dependent on the pH and temperature. These two parameters can affect the exchange rate drastically. One pH unit change can create a tenfold difference in the exchange rate. Even small structural changes can be monitored by this technique [17 - 20, 35]. One of the main considerations of this technique is to minimize back exchange of amides during work up. For this reason, quenching step temperature must be 0°C and pH ~2.5 [63, 64]. Since pepsin is active at low pH it is the choice of protease for HDX-MS experiments. For HDX-MS experiments, ESI-MS system is more suitable than MALDI–MS because it is possible to control temperature in ESI-MS system.

Knowing that solvent accessibility impacts the HDX rate, protein interaction with another protein, peptide, DNA or a small molecule can be monitored using this method [45, 46, 56]. When a protein interacts with a molecule, interacting surface will be less accessible to solvent thus, interacting region will slowly exchange as compared to apoprotein.

Global HDX-MS is usually performed before local HDX-MS to see whether there is change or not or how big is the change (bigger the mass difference means bigger the interaction area). Global HDX-MS can be divided into following steps:

- Initially, a mass spectrum of apoprotein in an H_2O containing buffer is taken (this also can be called time zero). Protein is then incubated with the same buffer prepared in D_2O at varying time points. This H to D exchange can be stopped using quenching solution, which is typically an acidic solution kept on ice (0.1-1.0% FA).
- In the second round of global HDX-MS experiments, at first, protein is interacted with an analyte of interest, which could be a peptide, protein, DNA or an organic compound. Then a mass spectrum is taken (this spectrum will be the same as apoprotein because non-covalent interactions cannot be observed with denaturing MS conditions and non-covalent interactions require non denaturing MS). Then sample mixture is incubated with the same buffer prepared in D_2O at varying time points and quenched using a quenching solution. Finally mass spectrum is taken for D_2O incubated samples.
- The mass difference between deuterated apoprotein and protein-analyte gives how many amino acids interact with the analyte. One mass difference corresponds to one amide hydrogen therefore, equals to one amino acid.

Local HDX-MS experiments are performed to identify which amino acids are involved in interaction with the analyte of interest. Local HDX-MS experiments can be divided into the following steps:

- Initially, apoprotein in H_2O containing buffer is digested into peptide fragments using pepsin. Then, ESI-MS/MS is performed as described in section 5.1.2 of this chapter and the sequences of peptide fragments are identified. Same experiment is performed for the apoprotein incubated with same buffer prepared in D_2O. The peptide fragments are then analyzed using ESI-MS for the D_2O incubated sample. Peptide fragments are identified comparing with the H_2O experiment.
- In the second round of local HDX-MS experiment, protein is interacted with the analyte of interest then HDX performed in D_2O containing buffer. After D_2O incubation protein-analyte mixture digested into peptide fragments by pepsin and then ESI-MS is performed to identify peptides.
- The mass difference of identified peptides between two sets of HDX-MS experiments provides protein analyte interaction at a structural level.

Important Notice: When performing HDX-MS experiments after quenching step samples have to be kept cold. Even HPLC solvents and HPLC column need to be placed on an ice bath to minimize the back exchange rate.

5.1.4.2. Chemical Cross Linking Coupled by MS (XL-MS)

In addition to HDX-MS experiments, another MS based method called chemical cross-linking coupled with MS (XL-MS; sometimes abbreviated as CX-MS) can be used to monitor protein-protein and protein-peptide interactions. Chemical Cross Linking as its name refers cross links two amino acid residues using an organic compound often called cross linker. A cross linker is a molecule that contains two or more reactive ends with the ability to attach functional groups. This technique is based on stabilizing an existing interaction *via* covalent binding using a cross linker [57 - 60, 69]. If the attachment occurs between two functional groups in a protein tertiary, structure gets stabilized.

There are many proposed cross linker molecules but only few of them are used for proteomics applications. The reason is that protein stability is very sensitive to pH, temperature and buffer conditions. Therefore crosslinking reaction should not affect protein stability, which limits the cross linker choice. Chemical cross-linking reactions target Lys, Asp and Glu functional groups in proteins [57 - 60]. The most common amine group reactive cross linkers used in proteomics are Disuccinimidyl substrate (DSS), disuccinimidyl glutarate (DSG) and Bis(sulfosuccinimidyl) substrate (BS3). One thing to keep in mind is that DSS initially is not soluble in water. First, it needs to be dissolved in an organic solvent like DMSO or DMF then it can be used. Buffer of choice for crosslinking reaction can be HEPES, phosphate or borate buffers. Carboxylic acid reactive cross linkers are   4-(4,6-dimethoxy-1,3,5-triazin-2-yl)-4-methylmorpholinium   chloride (DMTMM) and 1-ethyl-3-(3-dimethylaminopropyl) carbodiimide hydrochloride (EDC) [57 - 60]. EDC works at low pH values (~5), this may cause a problem, if protein of interest is not stable at low pHs.

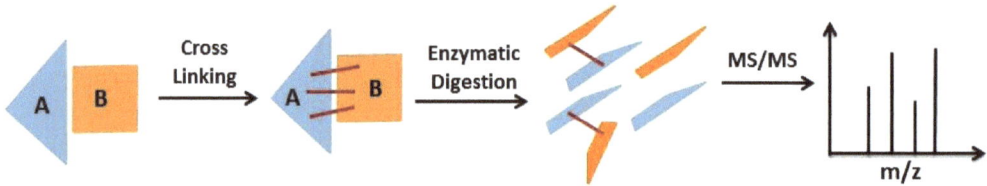

Fig. (14). Schematic diagram of XL-MS method. A and B are two interacting proteins or peptides.

Fig. (**14**) shows the workflow for applying XL-MS. XL-MS application can be divided into the following steps:

- Initially to stabilize the interaction between proteins or peptides of interests sample mixture is incubated with a cross linker. Incubation time can be from few minutes to a few hours depending on the protein and cross linker concentration. pH of the buffer is decided based on protein stability and cross linker working conditions.
- After adequate incubation, time attachment is checked using ESI-MS or MALDI-MS.
- In order to specifically determine where the interaction is taking place, sample mixture is digested into peptide fragments using trypsin, then MS/MS analysis is performed as described in 5.1.2 section of this chapter.

5.1.5. Protein Folding and Unfolding Rates

After proteins synthesized, they fold into a 3-dimentional structure based on an amino acid sequence. Protein folding is one of the most challenging issue in protein science. Proteins need to be properly folded to perform their biological functions. Sometimes proteins misfold, and these misfolded proteins are degraded in proteasome. Sometimes, misfolded proteins do not degraded and may form insoluble aggregates. These aggregates cause some serious diseases such as Mad Cow (prion misfolding), Alzheimer's disease (amyloid plagues) and Parkinson's disease (alpha-synuclein aggregates) [66, 70 - 72]. Today, understanding protein folding is a major research field in protein science. It is possible to determine protein folding and unfolding rates. There is a correlation between solution structure of a protein and the gas phase structure. A folded protein is more compact which leads to less z (higher m/z). When proteins unfolded, more z (lower m/z) is expected because an unfolded protein has more surface area for more charges.

Protein folding and unfolding rates can be determined using HDX-MS method [61, 67] (see section 5.1.4.1). HDX-MS is a very sensitive method and even very small conformational changes can be monitored. When proteins unfolded and/or formed aggregates, they show different HDX-MS profiles compared to properly folded protein. Change in HDX-MS profile indicates a conformational change.

5.2. Microbiology

MS is the main tool in proteomics laboratories since the beginning of 1980s. In recent years, the use of MS in microbiology is also becoming popular with the development of new methods. Usually, MALDI-TOF is used to identify a microorganism such as bacteria and fungi [36 - 39]. It allows a rapid and low-cost determination from broth or planted media.

An important consideration in bacterial identification is that the same species can have different mass spectral fingerprints depending on the growth conditions and harvesting solutions. Obtained spectrums are reproducible as long as bacteria are grown under the same conditions. The correct results also highly depend on sample preparation such as, choice of matrix and matrix solution affect the mass spectrum significantly. Bacterial identification is simply done by comparing mass spectral fingerprints that are obtained under identical growth and sample preparation conditions.

5.3. Medicine

Improvement of MS instrumentation and development of new methods allowed the use of MS in clinical laboratories. Clinical laboratories use MS for the diagnosis of disease, identification of metabolic disorders, discovering new biomarkers and identifying drug toxicity [40 - 44, 72 - 74]. It provides fast and low cast diagnosis of disease.

For example, serum steroids play a role in endocrine disorders, and by measuring steroid hormones in serum using LC-MS/MS, these disorders can be diagnosed. Some of the reported steroids that can be measured by LC-MS/MS are cortisol, estradiol, progesterone, corticosterone, androstenedione and 11-deoxycortisol [41]. Measuring these steroids with traditional immunoassays are time consuming and not as specific as LC-MS/MS measurements. In addition to steroid analysis, Vitamin D deficiency analysis by MS is also one of the established applications of MS in medical laboratories. 25-hydroxyvitamin D_3 is the most abundant and stable form of vitamin D in blood serum and its concentration in blood serum can diagnose its deficiency [41, 72].

MS can also be used to diagnose certain disease in newborns by determining the defects of amino acids, organic acids and fatty acids metabolism. These diagnostic tests include urinary amino acid analysis, plasma amino acid analysis and plasma acylcamitines analysis using MS/MS as reported in [41, 74]. MS applications in clinical laboratories still need development of new methods and/or improvement of existing methods.

CONSENT FOR PUBLICATION

Not applicable.

CONFLICT OF INTEREST

The author (editor) declares no conflict of interest, financial or otherwise.

ACKNOWLEDGEMENTS

Declared none.

REFERENCES

[1]   S.G. Roussis, and A.S. Cameron, "Simplified hydrocarbon compound type analysis using a dynamic batch inlet system coupled to a mass spectrometer", *Energy Fuels,* vol. 0624, no. 20, pp. 879-886, 1997.
[http://dx.doi.org/10.1021/ef960221j]

[2]   C.H. Evans, "Introduction to mass spectrometry", *Trends Biochem. Sci.,* vol. 11, no. 3, pp. 124-125, 1986.
[http://dx.doi.org/10.1016/0968-0004(86)90058-7]

[3]   T.R. Covey, E.D. Lee, A.P. Bruins, and J.D. Henion, "Liquid chromatography/mass spectrometry", *Anal. Chem.,* vol. 58, no. 14, pp. 1451A-1461A, 1986.
[http://dx.doi.org/10.1021/ac00127a001] [PMID: 3789400]

[4]   C. R. Blakley, J. J. Carmody, and M. L. Vestal, "A new soft ionization technique for mass spectrometry of complex molecules, " J. Am. Chem. Soc., vol. 920, no. 1976, pp. 5931–5933, 1980
[http://dx.doi.org/10.1021/ja00538a050]

[5]   M. Karas, and F. Hillenkamp, "Laser desorption ionization of proteins with molecular masses exceeding 10,000 daltons", *Anal. Chem.,* vol. 60, no. 20, pp. 2299-2301, 1988.
[http://dx.doi.org/10.1021/ac00171a028] [PMID: 3239801]

[6]   A.P. Bruins, "Mechanistic aspects of electrospray ionization", *J. Chromatogr. A,* vol. 794, no. 1–2, pp. 345-357, 1998.
[http://dx.doi.org/10.1016/S0021-9673(97)01110-2] [PMID: 9764504]

[7]   C.S. Ho, C.W. Lam, M.H. Chan, R.C. Cheung, L.K. Law, L.C. Lit, K.F. Ng, M.W. Suen, and H.L. Tai, "Electrospray ionisation mass spectrometry: principles and clinical applications", *Clin. Biochem. Rev.,* vol. 24, no. 1, pp. 3-12, 2003.
[PMID: 18568044]

[8]   M. Wilm, "Principles of electrospray ionization," Mol. Cell. Proteom., vol. 10, no. 7, p. M111.009407, 2011

[9]   S. Aldrich, "MALDI-Matrices : Properties and Requirements • Sample Preparation Techniques • Matrix Applications • MALDI-MS for Inorganics," AnalytiX, 2001

[10] P.M. Uthe, "Quadrupole mass analyzer" U.S. Patent No 3457404, 1969

[11] M. Kozo, "Quadrupole mass spectrometer," U.S. Patent No 5227629, 1993

[12] P.H. Dawson, "Quadrupole mass analyzers : Performance, design and some recent applications", *Mass Spectrom. Rev.,* vol. 5, no. 0, pp. 1-37, 1986.
      [http://dx.doi.org/10.1002/mas.1280050102]

[13] *Acc. Chem. Res.,* vol. 13, no. 2, pp. 33-39, 1980.
      [http://dx.doi.org/10.1021/ar50146a001]

[14] R.A. Yost, and D.D. Fetterolf, "Tandem mass spectrometry (MS/MS) Instrumentation", *Mass Spectrom. Rev.,* vol. 2, no. 1, pp. 1-45, 1983.
      [http://dx.doi.org/10.1002/mas.1280020102]

[15] A.I. Papayannopoulos, "The interpretation of collision☐induced dissociation tandem mass spectra of peptides", *Mass Spectrom. Rev.,* vol. 14, no. 1, pp. 49-73, 1995.
      [http://dx.doi.org/10.1002/mas.1280140104]

[16] A.M. Falick, W.M. Hines, K.F. Medzihradszky, M.A. Baldwin, and B.W. Gibson, "Low-mass ions produced from peptides by high-energy collision-induced dissociation in tandem mass spectrometry", *J. Am. Soc. Mass Spectrom.,* vol. 4, no. 11, pp. 882-893, 1993.
      [http://dx.doi.org/10.1016/1044-0305(93)87006-X] [PMID: 24227532]

[17] L.M. Mikesh, B. Ueberheide, A. Chi, J.J. Coon, J.E. Syka, J. Shabanowitz, and F.D. Hunt, *"The utility of ETD mass spectrometry in proteomic analysis* vol. 1764. Biochimica et Biophysica Acta (BBA)-Proteins and Proteomics, 2006, no. 12, pp. 1811-1822.

[18] M.S. Kim, and A. Pandey, "Electron transfer dissociation mass spectrometry in proteomics", *Proteomics,* vol. 12, no. 4-5, pp. 530-542, 2012.
      [http://dx.doi.org/10.1002/pmic.201100517] [PMID: 22246976]

[19] M. Sakairi, T. Mimura, T. Ishizuka, M. Tomioka, Y. Takada, and T. Nabeshima, T. Mimura, T. Ishizuka, M.Tomioka, Y. Takada, and T.Nabeshima. "Ion trap mass spectrometer," U.S. Patent No 6157030, 2000

[20] H-C. Chang, W-P. Peng, Y. Cai, and S-J. Kuo, W-P. Peng, Y. Cai, and S-J. Kuo. "Ion trap mass spectrometer," U.S. Patent No 6777673, 2004

[21] K. Yoshinari, "Ion trap mass spectrometer," U.S. Patent No 6,121,610, 2000

[22] Q. Hu, R.J. Noll, H. Li, A. Makarov, M. Hardman, and R. Graham Cooks, "The Orbitrap: a new mass spectrometer", *J. Mass Spectrom.,* vol. 40, no. 4, pp. 430-443, 2005.
      [http://dx.doi.org/10.1002/jms.856] [PMID: 15838939]

[23] R.A. Zubarev, and A. Makarov, "Orbitrap mass spectrometry", *Anal. Chem.,* vol. 85, no. 11, pp. 5288-5296, 2013.
      [http://dx.doi.org/10.1021/ac4001223] [PMID: 23590404]

[24] S. Michaela, and A. Makarov, "Orbitrap mass analyzer–overview and applications in proteomics", *Proteomics,* vol. 6, no. S2, pp. 16-21, 2006.
      [http://dx.doi.org/10.1002/pmic.200600528]

[25] G.M. Hieftje, "Time of Flight Mass Spectrometer," U.S. Patent No 5614711, 1997

[26] B. A. Mamyrin, "Laser assisted reflectron time-of-flight mass spectrometry," Inter. J. Mass Spectrom. Ion Proc., vol. 131, no. 1-19, 1994
      [http://dx.doi.org/10.1016/B978-0-444-81875-1.50004-5]

[27] Y. Yoshida, "Time of flight mass spectrometer," U.S. Patent No 4,625,112, 1986

[28] B.A. Mamyrin, V.I. Karataev, and D.V. Shmikk, "Time-of-flight mass spectrometer," U.S. Patent No 4,072,862, 1978

[29]   M.L. Muga, "Time-of-flight mass spectrometer," U.S. Patent No 4,458,149, 1984

[30]   B.A. Mamyrin, V.I. Karataev, D.V. Shmikk, and V.A. Zagulin, "The mass reflectron, a new nonmagnetic time-of-flight mass spectrometer with high resolution", *Sovi. J. Exp. Theor. Phy.,* vol. 37, p. 45, 1973.

[31]   D.M. Lubman, W.E. Bell, and M.N. Kronick, "Linear mass reflectron with a laser photoionization source for time-of-flight mass spectrometry", *Anal. Chem.,* vol. 55, no. 8, pp. 1437-1440, 1983. [http://dx.doi.org/10.1021/ac00259a062]

[32]   M.R. Wilkins, E. Gasteiger, A.A. Gooley, B.R. Herbert, M.P. Molloy, P.A. Binz, K. Ou, J.C. Sanchez, A. Bairoch, K.L. Williams, D.F. Hochstrasser, and R.M. Servet, "High-throughput mass spectrometric discovery of protein post-translational modifications", *J. Mol. Biol.,* vol. 289, no. 3, pp. 645-657, 1999. [http://dx.doi.org/10.1006/jmbi.1999.2794] [PMID: 10356335]

[33]   O.N. Jensen, "Modification-specific proteomics: characterization of post-translational modifications by mass spectrometry", *Curr. Opin. Chem. Biol.,* vol. 8, no. 1, pp. 33-41, 2004. [http://dx.doi.org/10.1016/j.cbpa.2003.12.009] [PMID: 15036154]

[34]   E.S. Witze, W.M. Old, K.A. Resing, and N.G. Ahn, "Mapping protein post-translational modifications with mass spectrometry", *Nat. Methods,* vol. 4, no. 10, pp. 798-806, 2007. [http://dx.doi.org/10.1038/nmeth1100] [PMID: 17901869]

[35]   M. Mann, and O.N. Jensen, "Proteomic analysis of post-translational modifications", *Nat. Biotechnol.,* vol. 21, no. 3, pp. 255-261, 2003. [http://dx.doi.org/10.1038/nbt0303-255] [PMID: 12610572]

[36]   J.O. Lay Jr, "MALDI-TOF mass spectrometry of bacteria", *Mass Spectrom. Rev.,* vol. 20, no. 4, pp. 172-194, 2001. [http://dx.doi.org/10.1002/mas.10003] [PMID: 11835305]

[37]   A. Wieser, L. Schneider, J. Jung, and S. Schubert, "MALDI-TOF MS in microbiological diagnostics-identification of microorganisms and beyond (mini review)", *Appl. Microbiol. Biotechnol.,* vol. 93, no. 3, pp. 965-974, 2012. [http://dx.doi.org/10.1007/s00253-011-3783-4] [PMID: 22198716]

[38]   E. Carbonnelle, C. Mesquita, E. Bille, N. Day, B. Dauphin, J.L. Beretti, A. Ferroni, L. Gutmann, and X. Nassif, "MALDI-TOF mass spectrometry tools for bacterial identification in clinical microbiology laboratory", *Clin. Biochem.,* vol. 44, no. 1, pp. 104-109, 2011. [http://dx.doi.org/10.1016/j.clinbiochem.2010.06.017] [PMID: 20620134]

[39]   R.D. Holland, J.G. Wilkes, F. Rafii, J.B. Sutherland, C.C. Persons, K.J. Voorhees, and J.O. Lay Jr, "Rapid identification of intact whole bacteria based on spectral patterns using matrix-assisted laser desorption/ionization with time-of-flight mass spectrometry", *Rapid Commun. Mass Spectrom.,* vol. 10, no. 10, pp. 1227-1232, 1996. [http://dx.doi.org/10.1002/(SICI)1097-0231(19960731)10:10<1227::AID-RCM659>3.0.CO;2-6] [PMID: 8759332]

[40]   Q. Meng, "Mass spectrometry applications in clinical diagnostics," J. Clin. Exp. Pathol. S, p. 4172, 2013 [http://dx.doi.org/10.4172/2161-0681.S6-e001]

[41]   M.S. Rashed, P.T. Ozand, M.P. Bucknall, and D. Little, "Diagnosis of inborn errors of metabolism from blood spots by acylcarnitines and amino acids profiling using automated electrospray tandem mass spectrometry", *Pediatr. Res.,* vol. 38, no. 3, pp. 324-331, 1995. [http://dx.doi.org/10.1203/00006450-199509000-00009] [PMID: 7494654]

[42]   F.T. Peters, "Recent advances of liquid chromatography-(tandem) mass spectrometry in clinical and forensic toxicology", *Clin. Biochem.,* vol. 44, no. 1, pp. 54-65, 2011. [http://dx.doi.org/10.1016/j.clinbiochem.2010.08.008] [PMID: 20709050]

[43]   E.P. Diamandis, "Mass spectrometry as a diagnostic and a cancer biomarker discovery tool:

opportunities and potential limitations", *Mol. Cell. Proteomics,* vol. 3, no. 4, pp. 367-378, 2004.
[http://dx.doi.org/10.1074/mcp.R400007-MCP200] [PMID: 14990683]

[44]   T. Guo, R.L. Taylor, R.J. Singh, and S.J. Soldin, "Simultaneous determination of 12 steroids by
        isotope dilution liquid chromatography-photospray ionization tandem mass spectrometry", *Clin. Chim.
        Acta,* vol. 372, no. 1-2, pp. 76-82, 2006.
        [http://dx.doi.org/10.1016/j.cca.2006.03.034] [PMID: 16707118]

[45]   A.J. Percy, M. Rey, K.M. Burns, and D.C. Schriemer, "Probing protein interactions with
        hydrogen/deuterium exchange and mass spectrometry-a review", *Anal. Chim. Acta,* vol. 721, pp. 7-21,
        2012.
        [http://dx.doi.org/10.1016/j.aca.2012.01.037] [PMID: 22405295]

[46]   V.A. Roberts, M.E. Pique, S. Hsu, S. Li, G. Slupphaug, R.P. Rambo, J.W. Jamison, T. Liu, J.H. Lee,
        J.A. Tainer, L.F. Ten Eyck, and V.L. Woods Jr, "Combining H/D exchange mass spectroscopy and
        computational docking reveals extended DNA-binding surface on uracil-DNA glycosylase", *Nucleic
        Acids Res.,* vol. 40, no. 13, pp. 6070-6081, 2012.
        [http://dx.doi.org/10.1093/nar/gks291] [PMID: 22492624]

[47]   T.E. Wales, and J.R. Engen, "Hydrogen exchange mass spectrometry for the analysis of protein
        dynamics", *Mass Spectrom. Rev.,* vol. 25, no. 1, pp. 158-170, 2006.
        [http://dx.doi.org/10.1002/mas.20064] [PMID: 16208684]

[48]   K. Biemann, and S.A. Martin, "Mass spectrometric determination of the amino acid sequence of
        peptides and proteins", *Mass Spectrom. Rev.,* vol. 6, no. 1, pp. 1-75, 1987.
        [http://dx.doi.org/10.1002/mas.1280060102]

[49]   N. Sepetov, O. Issakova, V. Krchnak, and M. Lebl, "Peptide sequencing using mass spectrometry,"
        U.S. Patent No 5,470,753, 1995

[50]   L.M. Mikesh, B. Ueberheide, A. Chi, J.J. Coon, J.E. Syka, J. Shabanowitz, and D.F. Hunt, ""The
        utility of ETD mass spectrometry in proteomic analysis," Biochimica et Biophysica Acta (BBA)-",
        Proteins and Proteomics, vol. 1764, no. 12, pp. 1811-1822, 2006.
        [http://dx.doi.org/10.1016/j.bbapap.2006.10.003]

[51]   A.I. Kaltashov, "Probing protein dynamics and function under native and mildly denaturing conditions
        with hydrogen exchange and mass spectrometry", *Inter. J. Mass. Spectrom.,* vol. 240, no. 3, pp. 249-
        259, 2005.

[52]   A.I. Kaltashov, ""Probing protein dynamics and function under native and mildly denaturing
        conditions with hydrogen exchange and mass spectrometry, " Inter", *J. Mass Spectrom.,* vol. 240, no.
        3, pp. 249-259, 2005.
        [http://dx.doi.org/10.1016/j.ijms.2004.09.021]

[53]   S. Pektas, and M.J. Knapp, "Substrate preference of the HIF-prolyl hydroxylase-2 (PHD2) and
        substrate-induced conformational change", *J. Inorg. Biochem.,* vol. 126, pp. 55-60, 2013.
        [http://dx.doi.org/10.1016/j.jinorgbio.2013.05.006] [PMID: 23787140]

[54]   K. Biemann, *Contributions of mass spectrometry to peptide and protein structure," Biomed. Environ.
        Mass Spectrom* vol. 16. , 1988, no. 1-12, pp. 99-111.

[55]   T.R. Covey, E. Huang, and J. Henion, "Protein sequencing by mass spectrometry." U.S. Patent No.
        5,952,653. 14 Sep. 1999

[56]   M.J. Chalmers, S.A. Busby, B.D. Pascal, Y. He, C.L. Hendrickson, A.G. Marshall, and P.R. Griffin,
        "Probing protein ligand interactions by automated hydrogen/deuterium exchange mass spectrometry",
        Anal. Chem., vol. 78, no. 4, pp. 1005-1014, 2006.
        [http://dx.doi.org/10.1021/ac051294f] [PMID: 16478090]

[57]   A. Leitner, L.A. Joachimiak, P. Unverdorben, T. Walzthoeni, J. Frydman, F. Förster, and R.
        Aebersold, "Chemical cross-linking/mass spectrometry targeting acidic residues in proteins and
        protein complexes", *Proc. Natl. Acad. Sci. USA,* vol. 111, no. 26, pp. 9455-9460, 2014.

[http://dx.doi.org/10.1073/pnas.1320298111] [PMID: 24938783]

[58]   A.N. Holding, "XL-MS: Protein cross-linking coupled with mass spectrometry", *Methods,* vol. 89, pp. 54-63, 2015.
[http://dx.doi.org/10.1016/j.ymeth.2015.06.010] [PMID: 26079926]

[59]   A. Leitner, T. Walzthoeni, and R. Aebersold, "Lysine-specific chemical cross-linking of protein complexes and identification of cross-linking sites using LC-MS/MS and the xQuest/xProphet software pipeline", *Nat. Protoc.,* vol. 9, no. 1, pp. 120-137, 2014.
[http://dx.doi.org/10.1038/nprot.2013.168] [PMID: 24356771]

[60]   A. Sinz, "Chemical cross-linking and mass spectrometry to map three-dimensional protein structures and protein-protein interactions", *Mass Spectrom. Rev.,* vol. 25, no. 4, pp. 663-682, 2006.
[http://dx.doi.org/10.1002/mas.20082] [PMID: 16477643]

[61]   J. Clarke, and L.S. Itzhaki, "Hydrogen exchange and protein folding", *Curr. Opin. Struct. Biol.,* vol. 8, no. 1, pp. 112-118, 1998.
[http://dx.doi.org/10.1016/S0959-440X(98)80018-3] [PMID: 9519304]

[62]   T.M. Raschke, and S. Marqusee, "Hydrogen exchange studies of protein structure", *Curr. Opin. Biotechnol.,* vol. 9, no. 1, pp. 80-86, 1998.
[http://dx.doi.org/10.1016/S0958-1669(98)80088-8] [PMID: 9503592]

[63]   Z. Zhang, and D.L. Smith, "Determination of amide hydrogen exchange by mass spectrometry: a new tool for protein structure elucidation", *Protein Sci.,* vol. 2, no. 4, pp. 522-531, 1993.
[http://dx.doi.org/10.1002/pro.5560020404] [PMID: 8390883]

[64]   J.R. Engen, "Analysis of protein conformation and dynamics by hydrogen/deuterium exchange MS", *Anal. Chem.,* vol. 81, no. 19, pp. 7870-7875, 2009.
[http://dx.doi.org/10.1021/ac901154s] [PMID: 19788312]

[65]   C.M. Dobson, " "Protein folding and misfolding,"", In: *Nature, insight review articles* vol. 426. , 2003, pp. 884-890.

[66]   M.M.G. Krishna, L. Hoang, Y. Lin, and S.W. Englander, "Hydrogen exchange methods to study protein folding", *Methods,* vol. 34, no. 1, pp. 51-64, 2004.
[http://dx.doi.org/10.1016/j.ymeth.2004.03.005] [PMID: 15283915]

[67]   D.M. Ferraro, N. Lazo, and A.D. Robertson, "EX1 hydrogen exchange and protein folding", *Biochemistry,* vol. 43, no. 3, pp. 587-594, 2004.
[http://dx.doi.org/10.1021/bi035943y] [PMID: 14730962]

[68]   J.W. Back, L. de Jong, A.O. Muijsers, and C.G. de Koster, "Chemical cross-linking and mass spectrometry for protein structural modeling", *J. Mol. Biol.,* vol. 331, no. 2, pp. 303-313, 2003.
[http://dx.doi.org/10.1016/S0022-2836(03)00721-6] [PMID: 12888339]

[69]   M. Stefani, C.M. Dobson, and J.H.P. Pro, "Protein aggregation and aggregate toxicity: new insights into protein folding, misfolding diseases and biological evolution", *J. Mol. Med. (Berl.),* vol. 81, no. 11, pp. 678-699, 2003.
[http://dx.doi.org/10.1007/s00109-003-0464-5] [PMID: 12942175]

[70]   P.J. Thomas, B.H. Qu, and P.L. Pedersen, "Defective protein folding as a basis of human disease", *Trends Biochem. Sci.,* vol. 20, no. 11, pp. 456-459, 1995.
[http://dx.doi.org/10.1016/S0968-0004(00)89100-8] [PMID: 8578588]

[71]   J.W. Kelly, "Commentary The environmental dependency of protein folding best explains prion and amyloid diseases", *Proc. Natl. Acad. Sci. USA,* vol. 95, no. 25, pp. 930-932, 1998.
[http://dx.doi.org/10.1073/pnas.95.3.930] [PMID: 9843941]

[72]   M. Gergov, B. Boucher, I. Ojanperä, and E. Vuori, "Toxicological screening of urine for drugs by liquid chromatography/time-of-flight mass spectrometry with automated target library search based on elemental formulas", *Rapid Commun. Mass Spectrom.,* vol. 15, no. 8, pp. 521-526, 2001.
[http://dx.doi.org/10.1002/rcm.260] [PMID: 11312500]

[73]   I. Ojanperä, M. Kolmonen, and A. Pelander, "Current use of high-resolution mass spectrometry in drug screening relevant to clinical and forensic toxicology and doping control", *Anal. Bioanal. Chem.,* vol. 403, no. 5, pp. 1203-1220, 2012.
[http://dx.doi.org/10.1007/s00216-012-5726-z] [PMID: 22302167]

[74]   B. Wilcken, V. Wiley, J. Hammond, and K. Carpenter, "Screening newborns for inborn errors of metabolism by tandem mass spectrometry", *N. Engl. J. Med.,* vol. 348, no. 23, pp. 2304-2312, 2003.
[http://dx.doi.org/10.1056/NEJMoa025225] [PMID: 12788994]

CHAPTER 2

Structural Elucidation of Macromolecules

Ana Luísa Carvalho, Teresa Santos-Silva, Maria João Romão, Eurico J. Cabrita and **Filipa Marcelo**[*]

UCIBIO-REQUIMTE, Departamento de Química, Faculdade de Ciências e Tecnologia, Universidade Nova de Lisboa, 2829-516, Caparica, Portugal

Abstract: While the structures of many proteins and nucleic acids are known and available in the Protein Data Bank, the folding and active sites of many fundamental macromolecules are yet to be elucidated. However, structure determination is only a small part of the story; to fully understand the diversity of functions that macromolecules have in living systems, interactions have also to be considered. Most drugs interact with macromolecular targets and the understanding of ligand-macromolecular recognition is fundamental in medicinal chemistry either from a purely structural viewpoint or by considering other important effects such as: ligand and binding site dynamics, distortion energies, solvent interactions and entropy. This chapter provides a comprehensive view on the several methods that are nowadays used to obtain structural information about macromolecules and their interactions with small ligands. The methods presented encompass different levels of characterization, either coarse information about the 3D shape or detailed structural data at the atomic level. Furthermore, relevance is given to the structural characterization of ligand binding events, either from the ligand or from the macromolecule viewpoint. Some experimental details behind X-ray crystallography, Nuclear Magnetic Resonance (NMR) and Small Angle X-ray Scattering (SAXS) are described, with examples and applications. An overview on Cryo-Electron Microscopy (CryoEM) is also presented. These are common state-of-the-art tools that truly complement each other and should be used in an integrative way.

Keywords: Cryo-EM, Drug design, Macromolecular structure, NMR, Protein-ligand interactions, SAXS, X-ray crystallography.

INTRODUCTION

Proteins and nucleic acids are key macromolecular players in living systems. Behaving in a precise way, they are nevertheless subject to malfunction.

---

[*] **Corresponding author Filipa Marcelo:** UCIBIO, REQUIMTE, Departamento de Química, Faculdade de Ciências e Tecnologia, Universidade Nova de Lisboa, 2829-516, Caparica, Portugal; Tel: +351212948300; E-mail: filipa.marcelo@fct.unl.pt

Yusuf Tutar (Ed.)

Understanding the atomic and molecular details of enzymatic reactions or simpler recognition processes, cannot be dissociated from the knowledge about the three-dimensional structure. This knowledge comprises the global macromolecular folding, shape modifications, quaternary arrangements and, fundamentally, the features of catalytic centers, in the presence of substrates and/or inhibitors. State-of-the-art methodologies are contributing to this knowledge. Mainly due to the atomic details it can provide, X-ray crystallography is frequently the method of choice to obtain the 3D structures of macromolecules. Besides, it is also powerful in the strategic design of potential drugs that fit and interact in the active sites of medically important proteins. X-ray crystallography comprises a series of techniques, part of a laborious and time-consuming process that nevertheless has benefited from an increasing computing power in the recent years. Modern molecular biology techniques, commercial formulations and instrumentation dedicated to crystallization have also contributed to the exponential growth in the elucidation of macromolecular structures by X-ray crystallography [1].

However, and in spite of all the advances made, the limiting step is still obtaining single crystals of adequate quality for X-ray diffraction. Once this is achieved in a reproducible way, obtaining the final 3D structure does not have many other bottlenecks. Nevertheless, a thorough knowledge about the theory is still necessary, so that the right decisions are made to get to the final and, hopefully, correct structure. Decisions involve choosing the adequate X-ray sources, radiation wavelengths and phasing methods. Validation of the final set of coordinates is also a crucial step in providing the users of crystal structures with the best model, resultant from the human (subjective) interpretation of the data. Although with a quite different experimental approach, Nuclear Magnetic Resonance (NMR) spectroscopy can be an alternative to X-ray crystallography to obtain the 3D structures of macromolecules in solution or in the solid state, with the striking advantage that no crystals are required. However, in spite of all the advances made in the last years, structure determination by NMR is still limited by the size of macromolecule. It is probably when applied to the characterization of molecular recognition events in solution that NMR spectroscopy reveals its great advantages over X-ray. NMR is able to monitor molecular interactions in solution at atomic detail permitting to deduce the key features that govern ligand-receptor interactions as well as to understand the dynamics of interaction, either from the ligand or the receptor viewpoint. This information is crucial to realize better the living systems and to develop pharmaceutically relevant applications. Likewise X-Ray crystallography, in the last decade NMR became an attractive technique to be used in drug discovery and optimization with paramount importance in medicinal chemistry [2 - 4].

Other methodologies have proven valuable due to the structural information provided, which often complements Crystallography and NMR methods. Small angle X-ray scattering (SAXS) is an emerging technique with great potential for the characterization of macromolecules in solution. Its application to protein-ligand adducts, conformational changes or large complexes is attracting the structural biologists' attention. In recent years, Cryo-Electron Microscopy (Cryo-EM) has become a very relevant technique for structure elucidation with increasingly better resolution of data. Although (still) limited to the study of large macromolecular assemblies, it is a powerful alternative to solve 3D structures, in cases that cannot be approached by X-ray crystallography or NMR methods.

Either in an independent or integrated approach, the knowledge about these methodologies and their potentialities is definitely valuable when the structural elucidation of macromolecules, isolated or in complex, is envisaged. Recognition determinants and mechanistic and kinetic properties can be explained, opening the way to molecular modifications, either in the macromolecules or in interesting medical drugs that influence the performance of biological systems.

2.1. Solving the 3D Structure of Proteins Using X-ray Crystallography

Far from being a simple technique, X-ray crystallography comprises a series of methods and involves several steps (Fig. **1**). Starting from a purified macromolecule, the first main step is to find a crystallization condition that originates single crystals. Afterwards, diffraction of X-rays provides the data that will lead to the model that best represents the structure of the protein (millions of copies of which build up the crystal).

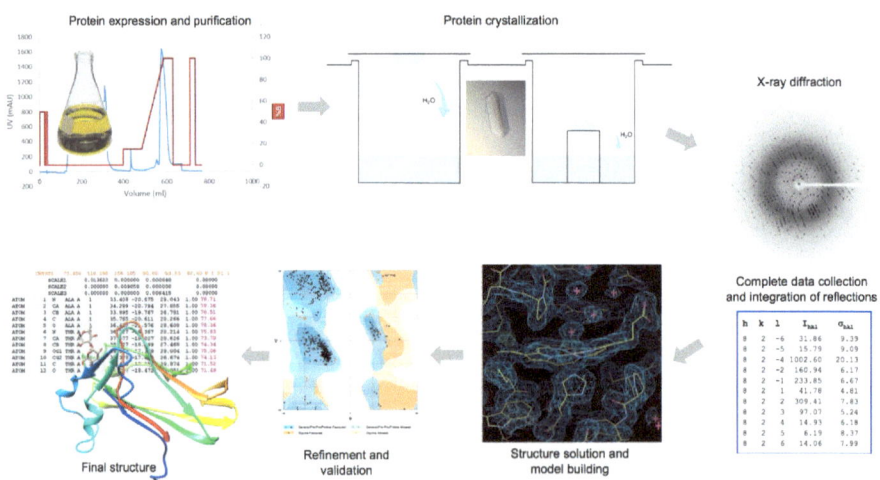

Fig. (1). Schematic overview of the main steps comprised in the determination of protein structures using X-ray crystallography, from the purified protein to the final three-dimensional structure.

X-ray photons are scattered by the electron clouds of the atoms and therefore the resulting "picture" after structure solution will be the electron density of the macromolecule. The atom coordinates that constitute the final model are built in this electron density, refined and validated before the usual deposition in the Protein Data Bank [5]. Presently, together with the atomic coordinates, all depositions require that structure factors also be deposited. These allow reconstructing the electron density and deriving further information about the macromolecular structure.

In the following sections, a general overview of macromolecular crystallography is presented, starting by crystallization, followed by the basic theory behind X-ray diffraction and structure solution methods and the criteria used to validate a structural model. Relevant details on particular subjects are included in the Notes section.

2.1.1. Protein Crystallization

The successful structure determination of proteins by X-ray crystallography is dependent on the availability of well-ordered single crystals. However, the production of good quality diffracting crystals is often the major bottleneck in protein crystallography in spite of the considerable progress and advances in the experimental methods. The use of fully automated crystallization robots is now possible in most labs and there is a large diversity of crystallization screenings commercially available.

To understand the difficulties encountered in crystallizing proteins, we first need to know what crystals are. Any crystal is characterized by an internal three-dimensionally ordered composition. Molecules in a crystal are packed and stabilized by intermolecular non-covalent interactions and the smallest packing motif (unit cell) is the smallest unit that can reproduce the whole crystal content by translations in the 3 dimensions of space (Fig. **2**).

In the case of crystals from biological macromolecules, those stabilizing interactions are complex and in fewer number than in crystals from small molecules (organic or inorganic) that usually crystallize easily. Protein crystals integrity is mainly due to ionic interactions or hydrogen bonds established between atoms from surface amino-acids and, very often, mediated by water molecules. The crystalline lattice is partially occupied by solvent molecules (40-60%) and is therefore little compact and with many cavities and channels (Fig. **3**).

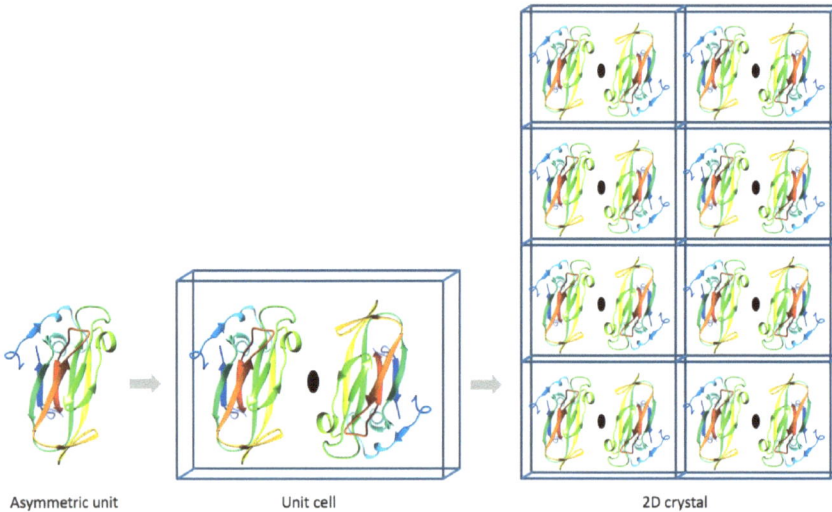

Asymmetric unit   Unit cell   2D crystal

Fig. (2). Assembly of unit cells to form a 2D crystal. In this simplified crystal, each unit cell contains 2 copies of the asymmetric unit packed according to the space group's symmetry operations (in this case, a single two-fold rotation). *a*, *b* and *c* are the cell axes; α, β and γ (in this case, all 90°) are the angles defined by b and c, a and c and, a and b, respectively. A protein crystal contains many copies of the unit cell propagated in the 3 directions of space.

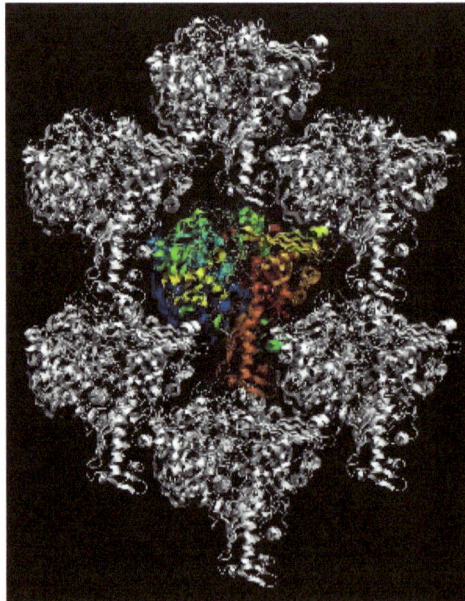

Fig. (3). Molecular packing inside a protein crystal revealing the intermolecular space. A protein crystal may contain from 40% to 60% solvent. The fragility of protein crystals is due to the large solvent channels, which permit the diffusion of small molecules. The protein depicted is the periplasmic aldehyde oxidase PaoABC from *Escherichia coli*, which crystallizes in space group *C*2 with unit cell parameters *a*=109.7 Å, *b*= 78.3 Å and *c*= 151.9Å, β= 99.7° (α= γ= 90°), and the crystals have a solvent content of 50%.

This is one of the reasons for the extreme mechanical fragility of protein crystals although it also constitutes an advantage for example for the preparation of protein-ligand complexes. In fact, due to the large solvent channels, it is possible to diffuse small molecules through the protein crystal lattice. This may allow for instance the binding of an enzyme inhibitor to the active site, producing an enzyme-inhibitor complex. This is a routine procedure in any drug development project (see below) [6].

The experimental methods for crystallizing proteins involve the slow achievement of a solubility minimum, corresponding to a certain degree of supersaturation. At this stage, small nuclei should start to appear and, in favorable conditions, protein molecules add to these nuclei, forming microcrystals (see Note 1). This process should occur very slowly to allow for molecules to be ordered in the crystal lattice and to promote crystal growth.

Standard methods for protein crystallization include varying a large number of experimental conditions, starting with pH, type and concentration of the precipitant and temperature. Very common precipitants are sulphates (NH_4^+ or Li^+), citrates and phosphates (Na^+ or K^+) as well as polymers like polyethylene glycol (PEG) of variable molecular weights (400 to 20000). Nowadays there is a large choice of commercial screenings that, in addition to precipitants, include a wide diversity of additives (ions and organic molecules). These comprise a variety of chemical compounds and small molecules whose purpose is to help stabilizing intermolecular crosslinks in protein crystals thus promoting lattice formation. There are numerous examples in the literature where additives were indispensable to enhance nucleation or crystal growth [7, 8].

The huge number of solutions available as well as the generalized use of automated crystallization systems (robots) allows thousands of conditions to be tested, provided that enough amount of pure protein is available. Typically, the concentration of the protein solution is ~5-15mg/ml, and drops are in the nanoliter range in the robot (or microliter when manually prepared).

In spite of several systematic studies regarding the influence of the multiple factors in protein crystallization, it is still difficult to make generalizations and to rationalize the factors that will determine the production of well-ordered single crystals of a certain protein.

Since crystallization is a very time consuming step and requires considerable amounts of protein, it is a good practice to use biophysical methods to assess the conformational stability and monodispersity of the protein sample. Being generally accepted that monodisperse (uniform) samples are more likely to crystallize, Dynamic Light Scattering (DLS) is often used to monitor changes in size and protein aggregation as a function of varying solution conditions [7]. Other equally common biophysical techniques used to assess the homogeneity and stability of the protein sample include Circular-Dichroism (CD) spectroscopy, Analytical Ultra Centrifugation (AUC) and above all, Differential Scanning Fluorimetry (DSF also known as thermofluor or thermal shift assay) [9, 10]. The latter was the first high-throughput thermal shift assay and is an inexpensive, simple and quick method and probes the refolding of a protein as it experiences progressive denaturation. It is currently the most popular method to evaluate the best conditions that may favor crystallization.

When the case of interest is the structure determination of a protein-ligand complex (inhibitor, substrate-analogue) as for example in structure-based drug design projects, additional crystallization experiments are necessary in order to obtain suitable crystals of the complex. Even in those cases where crystallization conditions are well established for the ligand-free protein, the formation of crystals of the co-structure may not be straightforward. Soaking and co-crystallization are the two most common methods used to prepare crystals of protein-ligand complexes (see review by McNae, *et al.* [11]). The soaking method is the simplest one since it involves only incubation of crystals of the unliganded-protein with the molecule of interest. Co-crystallization consists in preparing crystals using a solution of the protein-ligand complex. However, even when crystals are obtained and the crystal structure is solved, one very often finds that the ligand did not bind to the protein at all. These negative results are a waste of time and effort. To decrease the failure rate in obtaining crystals of the co-structure as much as possible, several biophysical methods should be used prior to co-crystallization in order to assess the experimental conditions that will most likely lead to complex formation. This preliminary analysis may increase the chances of obtaining the desired crystals, avoiding the time spent in "re-determining" unliganded-structures.

In addition to NMR (see section 2.2), other important biophysical methods employed to analyze ligand-protein binding in solution are Isothermal Titration Calorimetry (ITC), Surface Plasmon Resonance (SPR), Dynamic Light Scattering (DLS), Analytical Ultracentrifugation (AUC) and Differential Scanning Calorimetry (DSC) [7, 9, 12]. The use of one or more of these methods may allow to confirm ligand-protein binding and to estimate binding affinities.

2.1.2. Diffraction of X-rays by Protein Crystals

Once a single crystal of adequate size is obtained, it is subjected to X-rays and the produced diffraction is recorded. A crystal is built by many unit cells, that are described by three cell axes *a,b,c* and by three angles α, β and γ (Fig. **2**). The specific relationship between these six parameters defines the crystal's lattice and space group. Inside the unit cell, the asymmetric units (the macromolecules arranged with or without quaternary structure) are distributed according to the symmetry operations defined by the space group. The lattice and the space group will define the exact position of each spot (or hkl reflection) in the diffraction pattern produced by a crystal. h, k and l are also known as Miller indices.

A very important mathematical concept in crystallography, Bragg's Law, defines that the production of a spot in the diffraction pattern only occurs when constructive interference of the scattered radiation is attained (see Note 2). In practice and as the experimental result, the diffraction experiment produces a list of intensities (I_{hkl}) and associated errors for all reflections recorded. The intensity of each reflection results from the sum of photons that constitute the scattered wave, with an amplitude, phase and frequency (the same as the incident radiation). Historically recorded in photographic film, diffraction is nowadays detected by technologically advanced devices and computationally treated. Firstly, a few images are sufficient to predict the symmetry and unit cell parameters of the crystal (auto-indexing), as well as the orientation of the crystal relative to the beam. This is accompanied by refinement of several parameters including crystal, detector and beam parameters [13 - 16]. Once the unit cell parameters and space group are determined, it is possible to predict the amount of data that is required to measure all reflections (or as many as possible) produced by the crystal and a full data collection experiment is performed. After this, integration of all images takes place, extracting and listing the intensity and *hkl* position of each reflection, while also estimating an associated error. The high resolution limit of a diffraction data set is defined by the power of the crystal to divert the X-ray photons. The better the crystal, the further away from the beam center the diffraction spots will be detected and the higher the data resolution. Resolution is an important parameter for it defines how much detail can be included in the final structural model being highly dependent on the quality of the crystals (lattice disorder, mosaicity, protein flexibility, heterogeneities, among others).

2.1.3. 3D Structure Solution and Refinement

Each reflection spot in a diffraction pattern results from a monochromatic wave, constructively scattered by all equivalent lattice points. The intensity of the reflection corresponds to the intensity of this wave and is recorded by the detector. One other crucial piece of information to reconstruct the image of the macromolecule inside the crystal is the phase of the scattered wave when intercepted by the detector. This information cannot be retrieved and is in the loss of this phase angle that resides what is known as the Phase Problem in crystallography.

The Fourier Transform was adopted in crystallography, since it precisely describes the mathematical relationship between an object (the electron density) and its diffraction pattern (see Rupp, 2009 [17] for a comprehensive approach, not only, to crystallographic Fourier Transforms, but to the many other steps comprised in crystallography). Therefore, the intensities of reflections can be described by a Fourier series function of the electron density $\rho(x,y,z)$ distribution in the crystal and this is known as the structure factor equation, Eq. (1).

$$F_{hkl} = \int_V \rho(x,y,z) \cdot e^{[2\pi i(hx+ky+lz)]} \, dV \tag{1}$$

where F_{hkl} (the structure factor) is a vector defined by the amplitude, phase and frequency. This global structure factor corresponds to the vectorial sum of the scattering factors of all atoms that contributed to each *hkl* reflection. Applying the Fourier Transform to Eq. (1) allows obtaining the Electron Density equation, Eq. (2):

$$\rho(x,y,z) = (1/V) \sum_{hkl} |F_{hkl}| \cdot e^{2\pi i \alpha_{hkl}} \cdot e^{[-2\pi i(hx+ky+lz)]} \tag{2}$$

where α_{hkl} is the phase angle of reflection *hkl*, $|F_{hkl}|$ is the structure factor amplitude (proportional to $\sqrt{I_{hkl}}$), (x,y,z) are the fractional atomic coordinates in the unit cell and V is the volume of the unit cell. Using the Fourier transform and its inverse, the crystallographer can "travel" between the reciprocal space (the space of the diffraction pattern, the *hkl* reflections and the structure factors) and the real space (the space of the electron density and the atomic coordinates of the molecules inside the crystal).

I_{hkl}(and, consequently, $|F_{hkl}|$) is directly measured in the diffraction experiment, but phase angles need to be estimated. Solving the phase problem consists of correctly estimating these values, introduce them in the electron density equation and retrieve the electron density map that surrounds all the atoms of the macromolecule in the crystal.

There are 3 main methods to overcome the loss of the phase information and sometimes a combination of them is required. Access to high-intensity, tunable X-ray sources as the synchrotron is sometimes crucial for obtaining the phase information. Unlike the so called "in-house" sources that provide radiation of a fixed wavelength, synchrotron facilities offer the possibility to tune radiation to a wavelength of choice and crystallographers take advantage of the radiation absorption behavior of particular atoms in the crystal. The methods that rely on this behavior are known as Anomalous Dispersion methods: MAD (for Multi-wavelength Anomalous Dispersion) or SAD (for Single-wavelength Anomalous Dispersion). The anomalous dispersion phenomenon results from X-rays of a particular wavelength being absorbed by the inner electrons of heavy atoms in the crystal. Following absorption, X-rays are re-emitted after a certain delay, inducing a phase shift in all of the reflections, which can be detected by comparing datasets collected at different wavelengths [18]. The detected differences allow locating the heavy atoms and these atom positions serve as initial approximate phases. The heavy atoms can be 1)intrinsic to the protein (*e.g.* metalloproteins), 2) introduced by soaking compounds through the crystal's solvent channels (Fig. **3**) alternatively, proteins can be recombinantly expressed incorporating heavier atoms (*e.g.* selenium-replaced methionines or cysteines; see Note 3).

The possibility of soaking heavy atom compounds through a crystal's solvent channels is the basis of the second method for estimating initial phase angles: Multiple Isomorphous Replacement (MIR). The crystal that incorporated the heavy atom is named a derivative. The differences in the intensity of the reflections from a derivative crystal (when compared with the reflections from a native one) provide the location of the heavy atoms and these coordinates serve as initial phases. Usually, this method requires the preparation of several derivatives with different heavy atoms, and can be performed using an in-house X-ray equipment. If the diffraction data set is collected at the X-ray absorption wavelength of the heavy atom, this method can be combined with the anomalous dispersion method [19].

The third and more straightforward method, Molecular Replacement (MR), is probably the most used, a consequence of the increasing number of structures available in the PDB. This method requires that the atomic coordinates of a similar protein are available (previously determined by crystallography, NMR or even cryo-EM methods). Similarity is evaluated by alignment and comparison of the primary sequences of both proteins. In this method, the potentially similar model (sequence identity higher than 30%) is used to locate the protein of interest in the unit cell of the unknown structure. This is computationally performed in two steps, guided by the calculation of particular maps, known as Patterson maps (see Note 4): first, a rotation search finds the orientation of the molecule in the unit cell; second, a translation search throughout the unit cell finds the position of the previously rotated model. MR is particularly important for ligand binding studies, as the free structure can be used as a search model on a dataset of the crystal with the ligand bound.

Finding the correct initial phases makes it possible to calculate a first electron density map, followed by preliminary model building in the identifiable secondary structure features and/or other recognizable zones of the protein (Fig. **4**). Even with sophisticated software, model building can be tricky and is very much affected by the resolution of the diffraction data and the quality of the initial estimated phases.

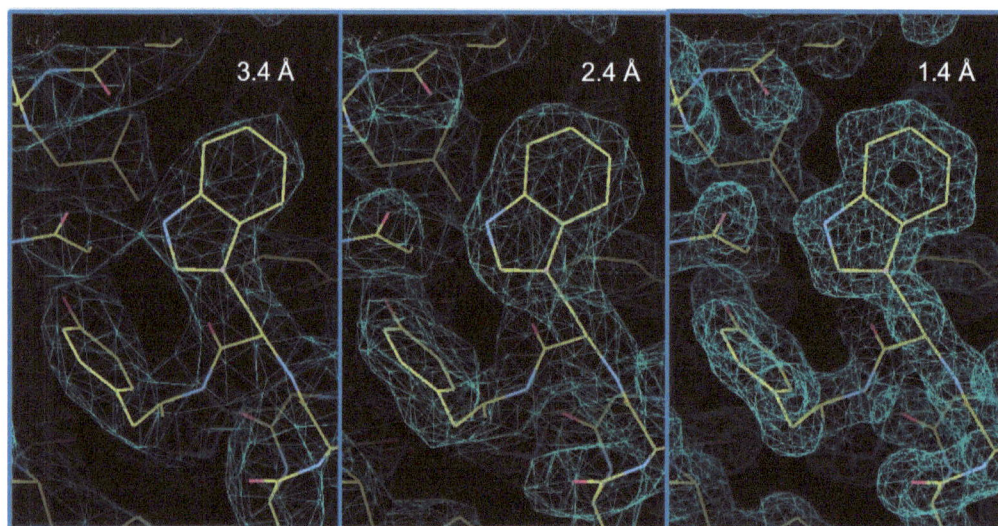

Fig. (4). Comparative view of an electron density map (contoured at 1 sigma level) calculated from data at 3 different high resolution limits (maps were calculated using FFT program implemented in the CCP4 suite [20] and displayed using COOT [21]).

Once correct phases are found, it is possible to calculate a preliminary electron density map. Initial model building can be performed "manually" in graphical software interfaces or automatically with autobuilding software. Generally, it involves a skeletonisation of the polypeptide chain along peaks of electron density and conversion of this skeleton in a polypeptide chain (main chain with branches of side chains) that complies with the stereochemistry of molecules (Engh & Huber parameterization and conformation-dependent stereochemical libraries) [22, 23]. Knowledge of the primary sequence is essential at this stage to identify and build stretches of sequence. After this, model refinement can take place adjusting positions of atoms as well as interatomic distances and angles in the electron density. This model will serve as a new set of phase angles, hopefully more accurate than the experimental phases. The phases from the model and the experimental phases can be combined and, together with the measured intensities (or amplitudes), are used to calculate a new electron density map (Eq. 2), where an improved and more complete model can be built (see Note 5 to learn more about electron density maps).

The file of coordinates that describes a macromolecular model solved by X-ray crystallography lists the 3 space coordinates (x, y and z) for each atom, its temperature factor or B-factor (Note 6) (a measure of how much an atom oscillates, iso- or anisotropically, around its position) and its occupancy (the fraction of molecules in which the atom is present throughout the crystal). A concept to have in mind is that all these parameters are refined for each atom against a relatively poor number of observations (the measured intensities). The average protein diffracts to a high-resolution limit around 2.0-3.0Å, where there is about one observation for each parameter. So, observations are needed and these are added to the model in the form of restraints and constraints [24]. Constraints are conditions intrinsic to the crystallized macromolecule and that cannot vary (*e.g.* occupancy of atoms and non-crystallographic symmetry, if present). Restraints are stereochemical impositions derived from the ideal values known for the high-resolution structures of small molecules. Examples of restraints include bond lengths, bond angles and close contact restrictions.

Besides revealing the folding of the macromolecular backbone and side chains, in every iteration of model building and refinement, the electron density can reveal many other molecules that have been orderly trapped in the crystalline lattice. In macromolecule-ligand studies, it is desirable that the electron density for the ligand of interest will be revealed and the model for that ligand should then be built and integrated in the list of coordinates. The level of detail observed for a ligand depends not only on the resolution or the quality of the diffraction data; affinity for a particular ligand of interest may influence the occupancy of the ligand and this reflects in the definition of the electron density. Also, although binding to the macromolecule, long ligands can exhibit partial disorder in regions that do not recognize the macromolecule and a partial model has to be built (Fig. **5**).

Other features usually present in the protein model are solvent molecules, mostly water molecules in ordered positions and forming a network around and inside the macromolecule. The number of observed solvent molecules is highly dependent on the maximum resolution of data and automated detection and addition of water molecules can be a great help in high resolution model building.

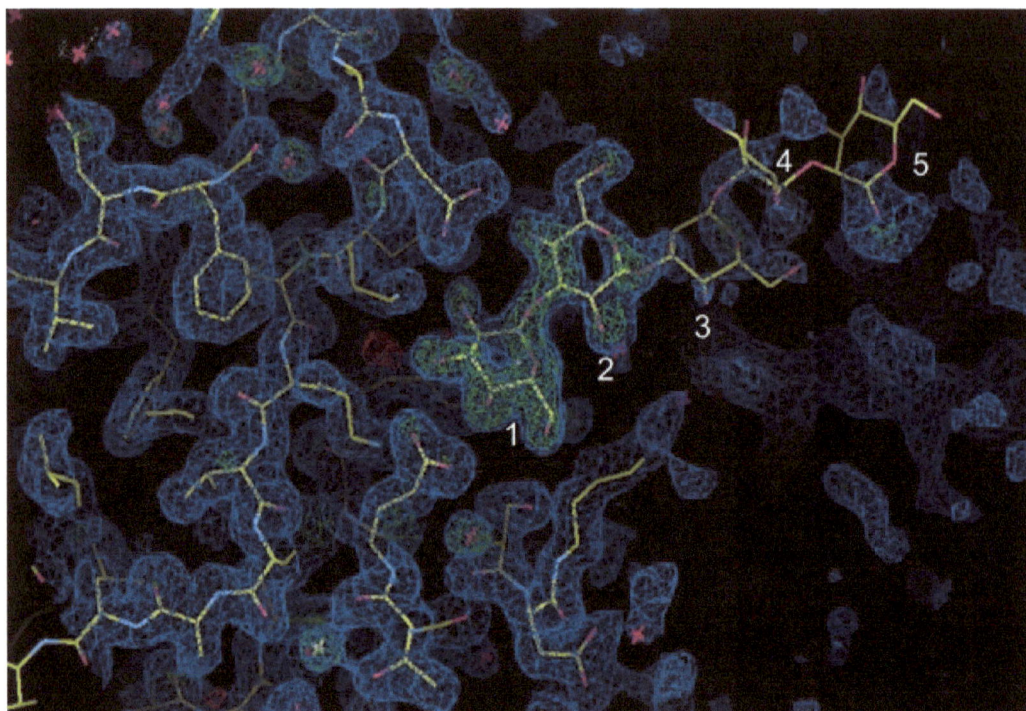

Fig. (5). $2F_{obs}$-F_{calc} electron density map (in blue and contoured at 1σ) calculated for a protein incubated with a pentasaccharide ligand prior to crystallization. The positive F_{obs}-F_{calc} map (in green, contoured at 3σ and calculated in the absence of the ligand) clearly reveals the position of the first 2 sugar units that are recognized by the protein amino acid side chains. Compared to these initial sugar units, the remaining three are more solvent-exposed and exhibit significant disorder. Iterative model building and refinement may improve phases for this zone, allowing construction of the complete pentasaccharide (unpublished). The picture was prepared using program Coot [21].

2.1.4. Validation and Analysis of Crystal Structures

The progress of model building and refinement is monitored by various validation criteria and these are crucial to prevent adding to the model more features than what is defined by the diffraction data. This is known as "over-fitting" and is a temptation that should be avoided. A widely used validation parameter is the R factor, which compares calculated structure factors $|F_{calc}|_{hkl}$, from the model, with observed structure factors $|F_{obs}|_{hkl}$, measured in the diffraction experiment, as shown in Eq. (3) [25]:

$$R = \frac{\sum(|F_{obs}| - |F_{calc}|)}{\sum|F_{obs}|} \qquad (3)$$

By itself, this parameter is not sufficient to detect over-fitting, because any random set of atoms added to the model will approximate F_{calc} to F_{obs} and lower the R factor. This is overcome by associating a cross-validation parameter, calculated in the same way, but using as F_{obs} only 5 to 10% of the unique reflections that are arbitrarily chosen and set apart from the normal refinement process. This is adequately known as the R_{free} factor and should not differ from the R factor by more than 5% [26]. Agreement with the limits of stereochemical restraints is also a validation criterion that accompanies model building and refinement cycles and is given in the form of root mean standard deviations (*rmsd*) for bond lengths and bond angles.

Misleading is thinking that the purpose of model building is to lower the R and R_{free} factors. Together with *rmsd* values, these parameters are good indicators of how well the model fits the data, but should not be seen as a target to achieve. The main purpose of model building is to find the model that best explains the measured data and hence the crystal's content. The final set of atom coordinates should reflect only the interpretable parts of the electron density calculated with the best possible set of phases, and this cannot be dissociated from the data quality and high-resolution limit.

Once refinement is taken to the best possible convergence, global validation takes place, evaluating several aspects, such as the distribution of amino acid residues in the energetically allowed regions of the Ramachandran plot (see Note 7), distribution of temperature factors (see Note 6), correctness of side-chain torsion angles, analysis of close contacts, water network contacts and other analysis, almost all implemented in software packages for validation (*e.g.* Molprobity [27]). This global validation will help correcting and finalizing the best structural model.

Open access policies guarantee the availability and continuous development of computer software packages (*e.g.* CCP4 [20] and Phenix [28]) dedicated to the calculations inherent to all here approached methods.

Structural analysis is done by "human" inspection of the model, literally visualizing atomic level interactions happening inside the model, but greatly assisted by specific software and platforms (*e.g.* wealth of tools, services and resources made available by the Protein Data Bank; http://www.ebi.ac.uk/pdbe/services). Such computational tools can reveal features with a biological and/or medical relevance. Intra- and intermolecular interactions can be thoroughly listed, protein-ligand relationships are computationally explored, and cavities with significant volume can be newly identified, as well as other unforeseen features that will inspire future questions and experiments.

No doubt, X-ray crystallography methods have benefitted from many recent advances. Access to tunable synchrotron microfocused sources, high sensitivity detectors, commercial crystallization screenings and tools, cryoprotection, recombination techniques to produce seleno-derivatized proteins, *etc.* are contributing to the increasing number of crystal structures available in the Protein Data Bank.

2.2. NMR Guide into Molecular Recognition

2.2.1. NMR and the Study of Intermolecular Interactions

It is perhaps in the study of intermolecular interactions that NMR spectroscopy reveals its great versatility and potential as an analytical technique for structural investigation. When applied to the study of dynamic processes, NMR spectroscopy allows obtaining structural information, with atomic resolution, related with the establishment or breaking of molecular interactions and in a variety of time scales. This is because NMR parameters are sensible to changes induced by processes in chemical exchange. Chemical exchange is defined as the transfer of atoms, molecules or molecular groups between two or more different regions or environments, being a frequent phenomenon in chemical, physical and biological systems. In NMR, these different chemical environments are typically associated with different resonance frequencies, coupling constants or relaxation times, characteristic of the regions between which the transfer takes place.

The simplest case of exchange is that of chemical shift exchange of a nucleus between two different environments, A and B, associated with the resonance frequencies, v_A and v_B, with equal populations (p_A and p_B). The relation between the frequency of the exchange process (k_{ex}, *i.e.* the lifetimes of the nucleus in the different sites) and the difference in frequency of the two environments ($v_A - v_B$) determines the NMR time-scale of the process, with consequences for the frequency related NMR observable, the chemical shift (δ). For a system in the slow chemical exchange limit ($k_{ex} \ll v_A - v_B$) the resonances of the nuclei in exchange are well resolved in the spectrum and two lines are observed at the "normal" chemical shifts of sites A and B. In the fast chemical exchange limit ($k_{ex} \gg v_A - v_B$) only one signal is observed, with its position defined by the weighted average of the populations of the two sites, $\delta_{obs} = p_A\delta_A + p_B\delta_B$. When going from the slow exchange to the fast exchange limit the resonance lines first broaden at their positions, δ_A and δ_B, entering an intermediate exchange regime until they almost disappear, then they coalesce to a single broad line at $\delta_{obs} = p_A\delta_A + p_B\delta_B$ when $k_{ex} \cong v_A - v_B$, and after that, as the ratio between k_{ex} and $v_A - v_B$ increases further the resulting line sharpens again, reaching the fast exchange limit when $k_{ex} \gg v_A - v_B$. In this context, in NMR, the effect of ligand binding can be monitored by perturbations observed on characteristic NMR parameters of either protein or ligand due to chemical exchange between the free and bound states. While changes in chemical shift between free and bound state are obvious perturbations to be explored, there are other useful NMR parameters affected by the exchange phenomena.

In solution, a large receptor and a small ligand have different motion properties, such as translational and rotational Brownian motion, the later associated to a specific correlation time (τ_c), that affect receptor and ligand NMR parameters such as, relaxation, diffusion and nuclear Overhauser effect (NOE) and make them completely distinct due to the difference in size. The correlation time, τ_c, is the average amount of the time a molecule takes to tumble through one radian and correlates with the molecular size with the relation $\tau_c \cong 10^{-12}$ Mw. Hence, on one hand, a large receptor, such as a protein, DNA or virus, has a large correlation time (ns-ms) inducing fast spin relaxation, a slow diffusion and negative and intense NOEs signals. On the other hand, a small ligand has a small correlation time (ps) which is related to slow relaxation, fast diffusion and positive and weak NOEs signals. Therefore, upon binding both ligand and receptor structures will be perturbed, and depending on the lifetime of the complex, which is related to the dissociation rate constant K_D, different ligand or receptor's NMR parameters will be more sensitive to the ligand-protein complex formation, thus providing a tool to monitor the interaction by solution NMR from a ligand or receptor-based approach (Fig. **6**-I). Currently, a repertoire of NMR methods is available to decipher ligand-receptor interactions from ligand- and receptor-based approaches. Furthermore, quantitative information about dissociation rate constants (K_D) can also be obtained by detecting either receptor or ligand resonances. Selection of the method will depend on the particular ligand-receptor complex under study, namely the off-rate of the dissociation process (k_{off}), receptor's size and of course the required structural information that is looked-for. The kinetics of the binding process, in particular the off-rate of the dissociation process is extremely relevant to select the method. Indeed, in fast-medium exchange conditions between the free and bound state ($k_{off} > 10$ s^{-1}), which correspond to transient to weak binding, NMR measurements to follow up ligand's parameters are more suitable enabling to disentangle ligand-receptor interactions (Fig. **6**-II). Alternatively, if the receptor's size allows, chemical shift and relaxation of specific protein resonances can be analysed in free and bound state. Receptor-detected methods are suitable in the cases of slow to intermediate binding regime ($k_{off} < 10$ s^{-1}) corresponding to weak to irreversible binding (Fig. **6**-II). This approach has the major advantage to give information on the residues directly involved in the recognition. However, these methods besides being limited by the size of the receptor require large amounts of selective isotopically labelled receptor and specific resonance assignment of the protein NMR spectra, a process that can be difficult and time consuming.

I

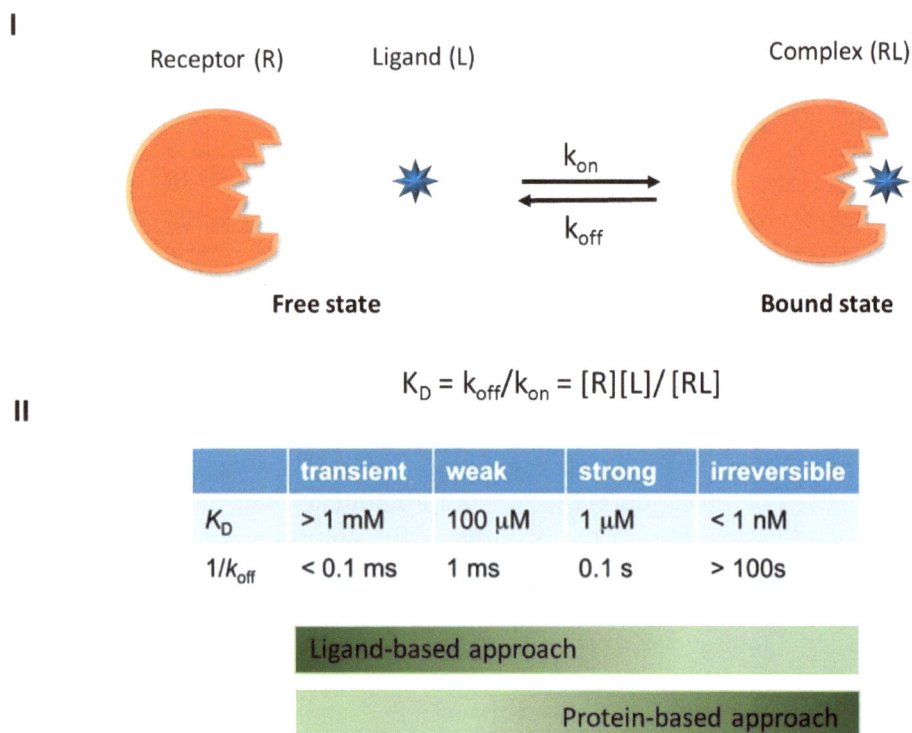

$$K_D = k_{off}/k_{on} = [R][L]/[RL]$$

II

	transient	weak	strong	irreversible
K_D	> 1 mM	100 μM	1 μM	< 1 nM
$1/k_{off}$	< 0.1 ms	1 ms	0.1 s	> 100s

Ligand-based approach

Protein-based approach

Fig. (6-I). Complex formation between a ligand and a large receptor with stoichiometry 1:1. The exchange rate of the association/dissociation equilibrium depends on the rate of the association k_{on} and dissociation k_{off} reactions with direct implications in the value of K_D where [R], [L] and [RL] are the equilibrium concentrations of receptor, ligand and complexed state, respectively. **II:** Assuming that k_{on} is diffusion controlled the choice of NMR-based approach strongly depends of k_{off} rate. Ligand-detected methods are more suitable for transient to medium binding while protein-based methods are more appropriate for strong to irreversible binding.

Herein, we described the most used and representative receptor and ligand-detected NMR based methods with examples of application and making special emphasis on their experimental aspects.

General aspects concerning to hardware and software technical aspects, as well as NMR sample preparation and basics in acquisition and procession NMR data are given in notes 8, 9 and 10.

2.2.2. Protein-detected NMR Based Experiments

The proton spectrum of a macromolecule is usually very complex due to many overlapped resonances as a consequence of the large number of nuclei in similar chemical environments. For the study of these complex systems it is usual to spread the ^{1}H NMR resonances in more dimensions combining it with the frequency of other nuclei (^{1}H, ^{15}N or ^{13}C) to reduce spectral overlap and aid in resonance assignment. In the case of proteins, the key spectrum is the heteronuclear correlation spectrum [^{15}N, ^{1}H]-HSQC, which can be considered the fingerprint of the protein (Fig. **7**). This 2D spectrum correlates N-H pairs, *i.e.* the frequency of hydrogen with that of an attached nitrogen. Therefore, in the [^{15}N, ^{1}H]-HSQC spectrum of a protein every backbone amide hydrogen will generate a peak, the dispersion of the peaks in the spectrum is a consequence of the different chemical environments generated by the three-dimensional structure of the protein. Chemical shift is very sensitive to structural changes therefore binding interactions will likely produce chemical shift perturbations. Monitoring ligand induced chemical shift perturbations (changes in peak intensity or position) in the [^{15}N, ^{1}H]-HSQC spectrum of a protein allows to quickly identify the binding interface if the assignment of the protein resonances is known as well as to obtain a value for the dissociation constant K_D [29]. Since ^{15}N is only 0.1% abundant, uniformly isotope labelling of the receptor is required for these receptor-based experiments. Depending on the type of protein this can be achieved by different strategies, but all of them rely on the production and purification of recombinant proteins and on the use of different compounds as sources of ^{15}N (or/and ^{13}C) [30].

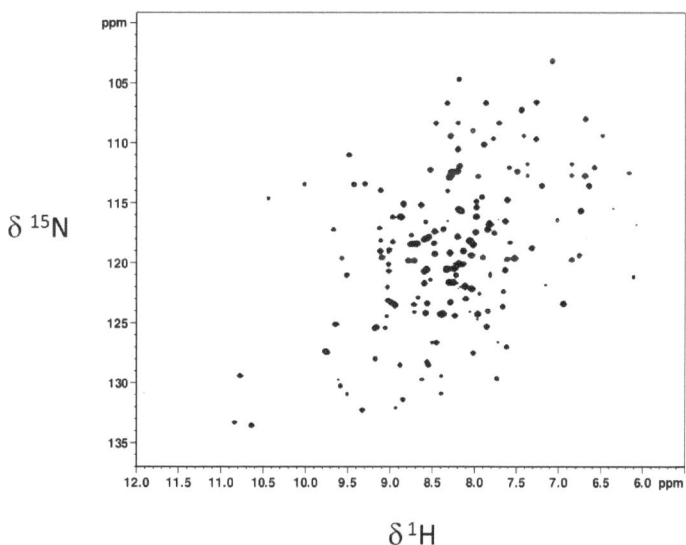

Fig. (7). The [^{15}N, ^{1}H]-HSQC spectrum of the carbohydrate binding module CBM11 (100 μM) from *Clostridium thermocellum* at 25^{0}C shows for every backbone amide hydrogen one cross peak at a typical position in the 2D map.

Since these methods depend on the observation of the protein signals they are usually limited by protein molecular weight, usually below 30 kDa, but using methods involving deuteration of the protein assisted with Transverse-Relaxation Optimized Spectroscopy (TROSY) techniques or selective residue labelling the upper limit can be extended to more than 100 kDa [31].

For sensitivity reason, rather large quantities of protein are also necessary, typically in the range of tens to hundreds of micromolar. When both the assignment of the signals in the [^{15}N, ^{1}H]-HSQC spectrum and the protein structure are known it is possible to widen the type of structural studies and explore other effects of the interaction with the ligand, such as detailed protein structural or dynamic changes, local or global. The methodology to perform the backbone assignment of the NH protein resonances in the [^{15}N, ^{1}H]-HSQC spectrum is a well-established practice. The most common relies in the use of triple resonance techniques with doubly labelled protein samples (^{13}C and ^{15}N) and in the analysis of a series of complimentary 3D NMR experiments that allow the sequential assignment of the residues. The NMR structure determination is mainly based on NMR data that allows the determination of interatomic distances and dihedral angles. This information is usually complemented with other NMR-derived restraints that help to obtain a more refined structure. Description of these procedures is however outside the scope of this chapter and for this purpose the reader is directed to the excellent detailed protocols already published [32, 33].

Another possibility for the mapping of protein interfaces uses the determination of amide-hydrogen exchange. If one adds deuterons to a protein (simply by dissolving a protein in D$_2$O) they normally exchange with the amide-hydrogens and the NMR signals vanish in the HSQC spectrum. How fast that happens depends whether an amide group is solvent exposed or not since good solvent accessibility allows a fast exchange. If a protein interacts with a ligand the amide-hydrogens at the interface should be shielded from the solvent molecules and slower exchange rates are expected for these residues when compared to the rates determined in the absence of ligand.

Also, NMR dynamics studies can yield information about the effect of ligand binding on the mobility of the involved side-chains and protein backbone at the interface, or even about dynamic changes in regions far away from the binding site. Measuring the relaxation rates R1 and R2 of ^{15}N, and steady state [^{15}N, ^{1}H]-NOE at different static magnetic fields and analyzing this data using the model-free approach [34, 35] in the presence and absence of the ligand delivers the parameters of molecular motion necessary for this analysis [36].

In the following section, we will present the chemical shift perturbation method, the most used and simple method for the identification of interaction sites.

2.2.2.1. Chemical Shift Perturbation

The primary NMR protein-based method for the detection of ligand-protein interactions relies on recording successive [^{15}N, ^{1}H]-HSQC spectra upon a titration of the protein with the ligand. Upon ligand binding the chemical environment of the protein backbone NHs changes due to the presence of the ligand and their chemical shift is altered, this change is usually more significant for those residues directly involved in the binding.

The type of chemical shift perturbation observed for the protein resonances in the [^{15}N, ^{1}H]-HSQC spectrum depends on the type of chemical exchange regime established between the free and the bound protein, which depends on the strength of the interaction with the ligand (Fig. **8**). When the exchange is slow in the chemical shift (that is for strong to tight binding) two sets of signals are observed in the spectrum, one corresponding to the resonances of the free protein and another to the resonances of the complex. As the concentration of ligand is raised the intensity of the signals corresponding to the free protein form will be reduced and those of the complex will increase. In these cases, the ligand-to-protein ratios necessary to observe protein chemical shift changes are rather low (bellow 1-2 equivalents) since the complex readily forms in solution and the identification of the perturbed residues is very clear. For a 1:1 binding the determination of an approximate K_D is also straightforward, since the intensity of the free and bound signals is proportional to the concentration of the free and bound protein.

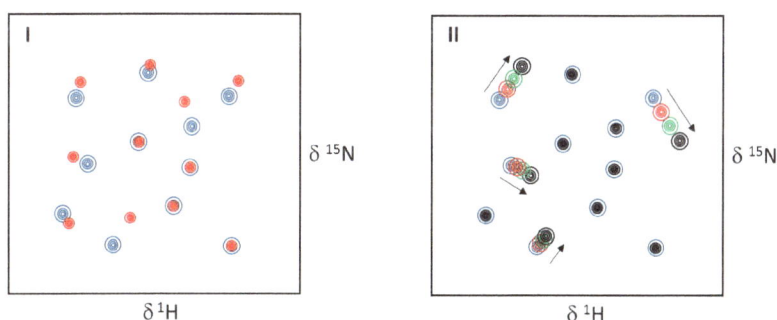

Fig. (8). Schematic representation of the chemical shift perturbation induced by the interaction of an unlabelled small ligand with a ^{15}N-labelled protein as detected in the [^{15}N, ^{1}H]-HSQC spectrum. **I.** Example of slow exchange (strong binding) after the addition of a sub-stoichiometric quantity of ligand; two signals are observed one for the free (blue) and another for the bound (red) protein. **II.** Example of fast exchange (weak binding), the scheme represents the superposition of 4 [^{15}N, ^{1}H]-HSQC spectra obtained for increasing quantities of ligand in the order blue, red, green and black; only one signal is observed in each spectrum with the shift (arrow) depending on the structural changes induced by the ligand at the specific residue.

In the fast exchange limit (weak to strong binding) only one resonance will be observed in the [^{15}N, ^{1}H]-HSQC spectrum, with its position determined by the population of free (p$_{free}$) and bound (p$_{bound}$) protein ($\delta_{obs} = p_{free}\delta_{free} + p_{bound}\delta_{bound}$). In the titration experiment, what will be observed is a shift of the δ_{obs} of the NH resonance in the 2D [^{15}N, ^{1}H]-HSQC spectrum due to the decrease of p$_{free}$ and increase in p$_{bound}$ as a consequence of ligand addition. Normally higher ligand-to-protein ratios are necessary to observe these shifts due to the higher K_D. In this regime, it is also possible to determine an approximate value for the K_D as will be illustrated in the next example.

Carbohydrate-binding modules (CBM) are non-catalytic protein modules from the cellulosome (a multi-subunit complex that associates a consortium of cellulolytic enzymes responsible for the degradation of the plant cell wall) that increase the efficiency of the catalytic modules. To identify the molecular determinants of carbohydrate recognition, the binding cleft of the family 11 carbohydrate binding module from *Clostridium thermocellum* (*Ct*CBM11) was identified by chemical shift perturbation [37] through protein titrations with cellotetraose and cellohexaose followed by [^{15}N, ^{1}H]-HSQC (Fig. **9**). Several NH amide pairs substantially change their chemical shift upon addition of increasing amounts of cellohexaose in order to better represent the distribution of affected and non-affected residues the combined chemical shift perturbation, $\Delta\delta_{comb}$, is determined for each titration point using the spectrum acquired in the absence of ligand as a reference for the initial chemical shit:

$$\Delta\delta_{comb} = \sqrt{(\Delta\delta_H)^2 + (0.1\Delta\delta_N)^2} \qquad (4)$$

where $\Delta\delta_H$ and $\Delta\delta_N$ are the differences in chemical shift determined for proton and nitrogen in relation to the [^{15}N, ^{1}H]-HSQC in the absence of ligand after each addition of ligand as shown in Eq. (4). Titrating ligand into protein so that the ligand eventually is in excess, thus saturating the protein binding site is the usual way to perform this study. In Fig. (**9**-I) the maximum $\Delta\delta_{comb}$ is represented for each residue of *Ct*CBM11. To determine if a residue belongs to the class of interacting residues a cut-off line is determined based on the standard deviation to zero [38]. The residues above this cut-off value are mapped into the protein structure allowing to identify the binding cleft, (Fig. **9**-II).

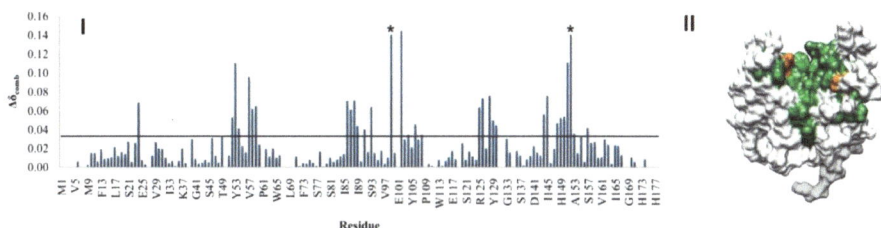

Fig. (9). Interaction of cellohexaose with CBM11 (100 µM) from *Clostridium thermocellum* at 25 °C. **I.** Identification of the amino acid residues that shift with the addition of cellohexaose by the representation of the Δδ_comb for each residue obtained at a saturating concentration of cellohexaose (2.0 eq.) **II.** Identification of the binding cleft of cellohexaose in the structure of *Ct*CBM11 by highlighting the residues above the cut-off line in (A).

The observed effects in the chemical shift indicate that the equilibrium is fast in the NMR time-scale. The interaction with cellohexaose undoubtedly shows that most changes occur for aromatic and charged residues (Tyr22, Tyr53, Tyr129, Tyr152, Asp99, Arg126, Asp128, Asp146) that can be involved in hydrogen bonding and aromatic stacking interactions with the oligosaccharide. Several of these residues were also identified as key for the binding process through site directed mutagenesis [39].

Based in the fact that the variations in chemical shift act as a marker for the binding equilibrium it is possible to use the combined chemical shift to estimate the dissociation constant (K_D) by adjusting the following equation to the experimental data where $[L]_0$ and $[R]_0$ are the total concentration of the ligand and the receptor in solution, respectively as depicted in Eq. (5) [40].

$$\Delta\delta_{comb} = \Delta\delta_{max} \frac{(K_D + [L]_0 + [R]_0) - \sqrt{(K_D + [L]_0 + [R]_0)^2 - (4[L]_0[R]_0)}}{2[R]_0} \quad (5)$$

In this case using Tyr129 a K_D of 19.2 µM was estimated.

There are many examples of the application of the chemical shift perturbation technique in the context of drug design, either for the screening of bioactive compounds or as a source of data to perform structure activity relationships by NMR. A recent and simple example of the use of chemical shift perturbation methods is reported in a project related to the development of new antibiotics for the characterization of the interactions between lipossacharides from the outer membrane (OM) of *Pseudomonas aeruginosa* with the outer membrane protein H (OprH) [41]. This interaction contributes to the integrity of the OM of *P. aeru-*

ginosa under low divalent cation and antibiotic stress conditions. Therefore, chemical shift methods are a useful tool to screen antibiotics that might disrupt OprH–LPS interactions and thereby increase the permeability of the OM of *P. aeruginosa*. Basics expects of ^{15}N-^{1}H-[HSQC] were included in note 11.

2.2.3. Ligand-detected NMR Based Experiments

Ligand-based NMR techniques can also be exploited to investigate molecular recognition events. In fact, ligand-protein binding can be deduced based on changes in the motion, orientation and diffusion properties of the ligands when passing from being free in solution to being recognized by a large receptor. These techniques have the benefit to require only small amounts of the receptor and, mainly they do not need selective isotopic labelling of the macromolecule. However, with ligand-observed methods, no indication of the protein binding site can be directly extracted. For that purpose, competition binding experiments, using a well-known binder, should be accomplished in order to deduce the binding site. Ligand-based methods are particularly useful to monitor ligands in the intermediate to fast off-rates between the free and bound states which corresponds to an equilibrium dissociation constant from mM to µM range of K_D. In the case of strong binders observation and analysis of the receptor signals are more suitable. The primary NMR methods for detection of ligand-protein interactions rely on measurement of T1/T2 relaxation rates of ligand resonances. Upon binding the molecular motion of the ligand dramatically changes and becomes similar to that of the receptor and selective shortening the T1/T2 relaxation times of ligand resonances in closer contact with the receptor. Reduction of T1/T2 relaxation times induces line broadening of certain protons of the ligand; some of them can even disappear. The binding of cellohexaose to the carbohydrate binding module *Ct*CBM11 was estimated by line broadening studies (T2 relaxation) [42].

Fig. (**10**) depicts that protons H6 and H2 from the central units of glucose are those that experienced a significant line broadening, which indicates their proximity with the CtCBM11 binding site. Alternatively, monitoring T1 values of galactose, in free and bound to human galectin-1 (hGal-1), highlighted a selective decrease of T1 value of the anomeric proton of galactose H1-Gal indicating ligand binding [43].

Nowadays, nuclear Overhauser effect (NOE) based techniques are the most robust techniques to detect biomolecular interactions. TrNOE (transferred-nuclear Overhauser effect) experiments allow to determine the ligand bioactive conformation while saturation transfer difference spectroscopy (STD-NMR) permits to decipher which protons are in close contact with the protein permitting to define the ligand epitope map. In addition, measurement of the diffusion coefficient of ligands in free and bound states by NMR diffusion ordered spectroscopy (DOSY) can be also employed to detect ligand-protein complexes in solution. In the following section, we will describe briefly the fundamentals of NOE and introduce with examples the trNOESY and STD-NMR experiments as tools to investigate ligand-protein interactions. At the end of the section we will report diffusion NMR methods applied to the molecular recognition field.

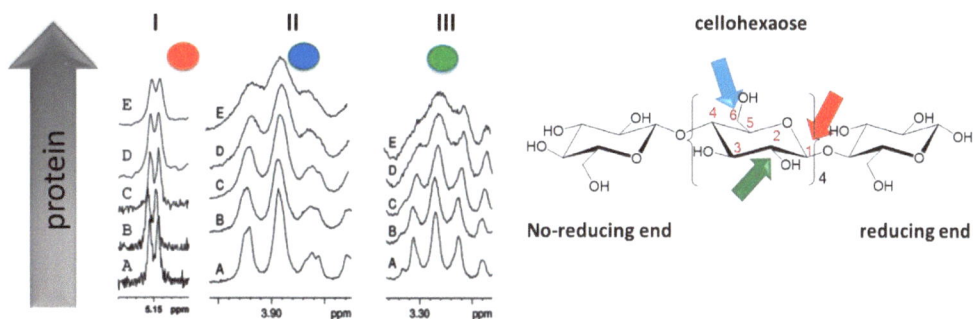

Fig. (10). Line broadening studies. **I-III.** Series of spectral regions of a solution of cellohexaose 0.787 mM in D$_2$O, corresponding to protons 1, 6 and 2 of cellohexaose, respectively, acquired at 298 K as a function of CtCBM11 concentration (A, 0.0 mM; B, 0.031 mM; C, 0.060 mM; D, 0.116 mM; E, 0.168 mM).

2.2.3.1. NOE-based Methods

The Nuclear Overhauser Effect (NOE) is defined by the change in intensity of one spin resonance due to the perturbation in spin transitions, between α and β states, of another dipolar coupled resonance. Consider two homonuclear spin-1/2 nuclei, I and S that share a dipolar (through space) coupling but that are not scalar coupled ($J_{IS} = 0$). The energy level diagram for this system is represented in Fig. **(11)**.

The perturbation in the equilibrium populations of spin S is achieved by either saturating a resonance, *i.e.*, equalizing the spin population differences, between α and β states, across the corresponding transitions of spin S (steady-state NOE), or inverting the population differences across the transitions of spin S (transient NOE). Then, to restore the equilibrium populations of spin S distinct mechanisms of relaxation, can occur. The single quantum transition, W1, corresponds only to the spin flip of spin S and does not affect the difference in populations of spin I. In contrast, the double and zero quantum transitions, W2 and W0 respectively, directly affect the difference in populations of spin I due to the cross-relaxation. In this context, W2 increases the difference in the populations between α and β states of spin I and result in the raise of signal intensity of spin I (positive NOE).

On the other hand, W0 decreases the difference in the populations between α and β states of spin I and will result in the decrease of signal intensity of spin I (negative NOE). The probability to occur W2 or W0 transitions will modulate the sign and intensity of NOE. The magnitude of the NOE detected for spin I when spin S is perturbed ($\eta_I\{S\}$) is expressed by Eq.(6), where I_0 corresponds to the intensity/integral of signal I in a normal spectrum and I corresponds to the intensity of signal I in the spectrum in which signal S is perturbed.

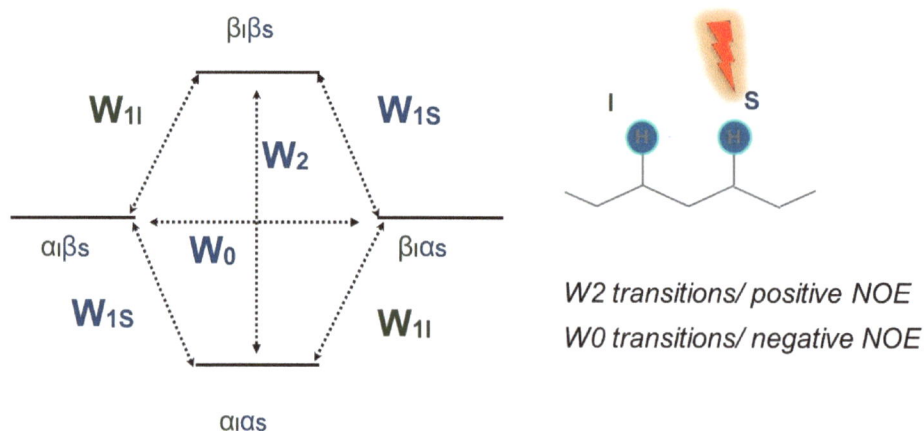

Fig. (11). Energy level diagram for a two homonuclear ½ nuclei spin system, I and S. The diagram displays the transition probabilities and spin states, showing the individual single quantum transitions W1I and W1S, and double and zero quantum transitions, W2 and W0, respectively. Only W2 and W0 are responsible for the sign of NOE.

$$\eta I\{I\} = \frac{I - I_0}{I_0} \times 100 \qquad \qquad (6)$$

Noteworthy, the probability to induce either W2 or W0 type of transitions depends of the molecular motion of the nucleus/molecule. In this sense, small molecules that tumble fast (small τ_c) yield positive and weak NOEs, while large molecules tumble slowly, (large τ_c) induce negative and strong NOEs (Fig. **12**). For molecules with intermediate size, there is an equal probability for W2 and W0 to occur and NOE approaches zero. The intensity of NOE depends on the relative orientation and the internuclear distance between I and S (Fig. **12**-I). In fact, NOE intensity has an inverse-sixth relationship with the distance, NOE $\propto r_{IS}^{-6}$. This relationship is the main value of NOE experiment since it permits to estimate the distance between protons and therefore deduce the conformation of molecules at different levels of complexity [44]. This correlation also means that the NOE falls away very rapidly with distance (observable NOEs < 5 Å). Finally, the cross-relaxation mechanism, essential for the NOE effect to occur, requires an amount of time to evolve, the so-called mixing time or saturation time of the experiment, which is defined depending of the molecule/system under study. For large molecules, the cross-relaxation or transfer of magnetization is very efficient and for that reason large molecules require lower values of mixing time. In the following two sections, we will describe NOE procedures employed to study ligand-protein interactions namely for ligands that are in fast association/dissociation equilibrium with a large receptor (*e.g.*, protein).

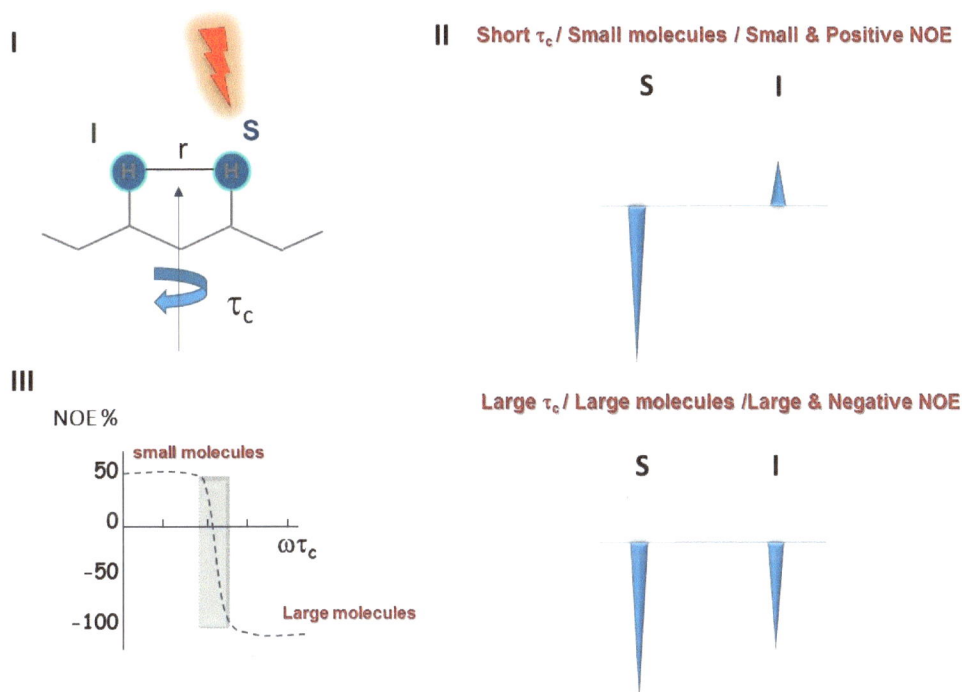

Fig. (12). I. The magnitude and sign of the NOE between I and S is dependent to the r^{-6} distance of the two nuclei and the correlation time τ_c of the molecule. **II-III.** Small molecules with short τ_c present small and positive NOEs while large molecules encode large and negative NOEs.

Transferred NOE

Upon binding, a small ligand adopts the molecular motion of the large ligand-protein complex and therefore the NOEs of ligand generated in the bound state change from positive to negative, and are named transferred NOEs (Fig. **13**) [45, 46]. The trNOE experiment is convenient for ligands in fast exchange between the free and protein-bound states since the NOEs generated in the bound state are only detected in the free state. Ligands with a slow off-rate in the relaxation timescale will not generate observable trNOE, since the relaxation of the ligand's signals will occur faster than the dissociation of the ligand from the receptor, thus preventing its detection. Moreover, if the equilibrium exchange rate is fast in the chemical shift timescale, due to the excess of the ligand the trNOEs will appear at the same chemical shifts of the free ligand which is very convenient for the assignment. The analysis of intramolecular NOEs in the trNOE spectrum allows to determine the bioactive conformation of the ligand. In certain cases, distinguishable intermolecular NOEs (ligand-receptor) can be detected allowing to determine the orientation of the ligand with respect to the binding pocket. The knowledge of the bound conformation of a small ligand is of paramount importance for rational based drug design and optimization. Basics aspects about sample preparation to record a trNOESY experiment in given in note 12.

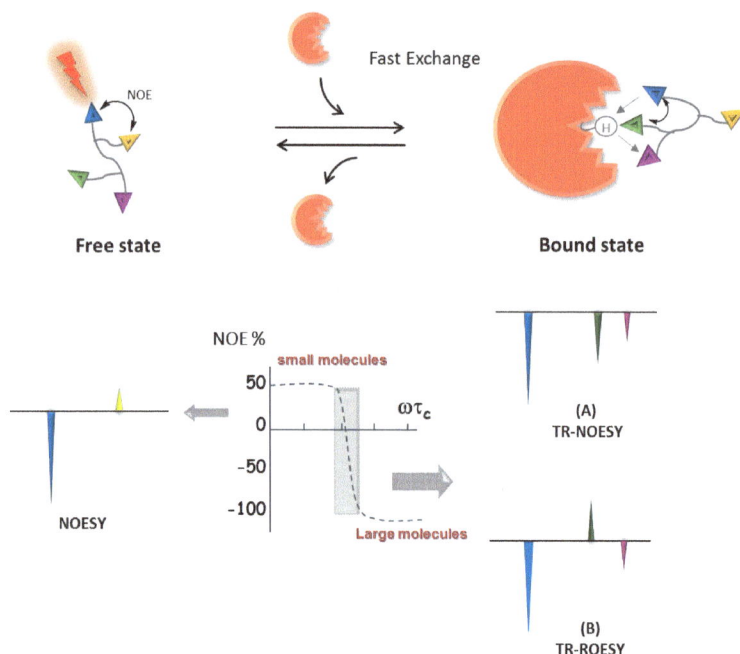

Fig. (13). The schematic view of the trNOESY experiment applied to study the geometry of a small molecule in fast exchange between free and bound state. Upon binding the positive NOE peaks of a small molecule will become negative (A), due to the increase of the correlation time. The analysis of the negative peaks encode information about the bioactive conformation of the ligand. The possible existence of spin diffusion effects can arise to indirect trNOE peaks complicating the analysis of trNOESY data. In this case, trROESY experiments (B) should be employed, allowing to discriminate between direct NOE signs to spin-diffusion mediated NOEs.

An experimental difficulty characteristic to trNOE measurements is related to spin diffusion issues. As a result, NOE signals may appear between two protons of the ligand, which are not actually close in space, but close to one or more protons within the ligand or at receptor (indirect interactions). In these cases, spin-diffusion mediated cross-peaks can be discriminated by employing trROESY experiments. In a trROESY experiment, direct NOE effects are positive while indirect NOEs are negative (Fig. **13**). Thus, trROESY allows to distinguish direct and indirect NOE effects.

Several applications of trNOESY have been described especially focused on the study of the structural features that govern glycan-protein interactions [47, 48]. In most of the cases, the ligand NOE spectra in the free and bound state are identical, highlighting that the ligand bioactive conformation is close to a specific conformation of the ligand already present in solution [49, 50]. This condition is thermodynamically favourable since it reduces the entropic cost of the binding process. However, there are other examples in which the flexibility of the receptor and/or ligand permits conformational changes of ligand upon binding. By combining NMR techniques, namely trNOESY, with biochemical and molecular modeling we investigated the global geometry of distinct C8-substituted guanine nucleotide inhibitors bound to the bacterial cell division protein FtsZ [51]. FtsZ, is a self-assembling GTPase protein, recognized as a target for new antibiotics [52]. The results pointed out changes in the recognition of C8-substituted guanine nucleotide inhibitors (8-pyrrolidino and 8-morpholino-analogue) by FtsZ from *Bacillus subtilis* (Bs-FtsZ). In free state both inhibitors present the *syn*-conformation around the glycosidic linkage. In the bound state, 8-pyrrolidino analogue binds in the *syn* conformation while 8-morpholino binds in the anti-conformation indicating a major conformational change between free and bound of morpholino analogue upon binding (Fig. **14**). This conformational switch may reduce the binding affinities of morpholino derivatives.

Also trNOESY of rosmarinic acid in presence of acetylcholinesterase (AChE) enabled to detect conformational distortion of the molecule, from an extended to a hairpin-like conformation [53]. Indeed, trNOESY experiment allowed to identify this molecule as the active component in *Salvia sclareoides* extracts. The ligand bioactive conformation can also be evaluated by means of half-filtered NOESY experiments. For ^{13}C-labelled proteins, NOESY ^{13}C-filtered experiments can be employed permitting to detect only the ligand protons by removing the cross-peaks attached to ^{13}C. This methodology was applied to determine the bound conformation of antithrombin–heparin pentasaccharide to the extracellular Ig2 domain of the fibroblast growth factor FGF1 and corresponding receptor (FGFR2) [54, 55].

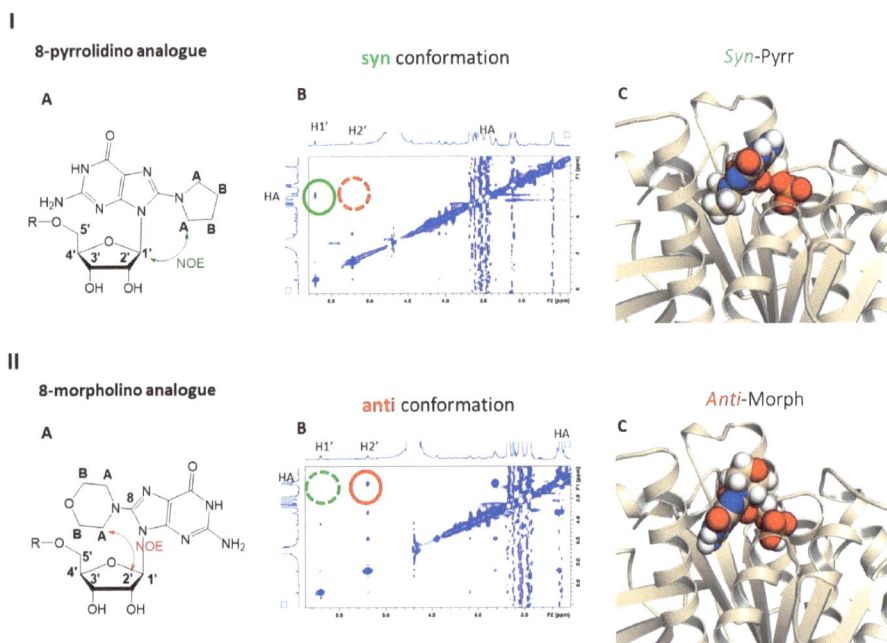

Fig. (14). Bound conformation of C8-substituted nucleotides in presence of *Bs*-FtsZ deduced by trNOESY experiments. **I.** 8-pyrrolidino analogue; **II.** 8-morpholino analogue. **A.** Chemical structure with numbering. **B.** trNOESY ligand/protein ratio 20:1, T=298K, 100ms mixing time. **C.** 3D binding modes of C8-substituted nucleotide in *Bs*-FtsZ nucleotide site.

Finally, the relative orientation of two competitive ligands A and B weakly bound to a common target can be achieved with the INPHARMA method [56]. This method consists to detect the inter ligand spin-diffusion mediated transferred-NOE data and provide relevant information for structure-based drug design with particular interest to the pharmaceutical industry.

Saturation Transfer Difference

Saturation transfer difference (STD-NMR) experiment is based on intermolecular NOE transfer from the large receptor to a interacting ligand (Fig. **15**) [57]. In this method, two ^{1}H-NMR spectra are accomplished, the off- and on-resonance spectrum, without and with selective saturation (selective r.f. pulse) of the receptor protons, respectively (Fig. **15**). If the ligand interacts with the receptor the intermolecular transfer of magnetization *via* ^{1}H-^{1}H cross-relaxation from the receptor to the bound ligand leads to a decrease of the ligand resonance intensities in the on-resonance spectrum (negative intermolecular NOE typical for large molecules) (Fig. **15**). This decrease can be better noticed in a difference spectrum that arises from subtraction of on-resonance spectrum to that spectrum recorded without saturation of the protein (off-resonance spectrum). Thus, the STD spectrum only displays the ligand signals that were in direct contact with the saturated protein resonances. Any ligand that is not interacting with the receptor does not receive any saturation from the protein and, therefore its intensities are not affected (Fig. **15**). The amount of transference of magnetization from the receptor to the ligand is proportional to the inverse of the sixth power of the distance between the ligand nuclei and receptor's protons. Therefore, the signal intensities of the protons in the STD spectrum relates to their proximity to the protons of receptor's binding site, allowing to quantify the individual contributions of each ligand's protons to the recognition event the so called ligand epitope mapping [58]. Usually, STD intensities obtained for a ligand are expressed as relative STD percentages (%) by normalizing to 100% the most intense one.

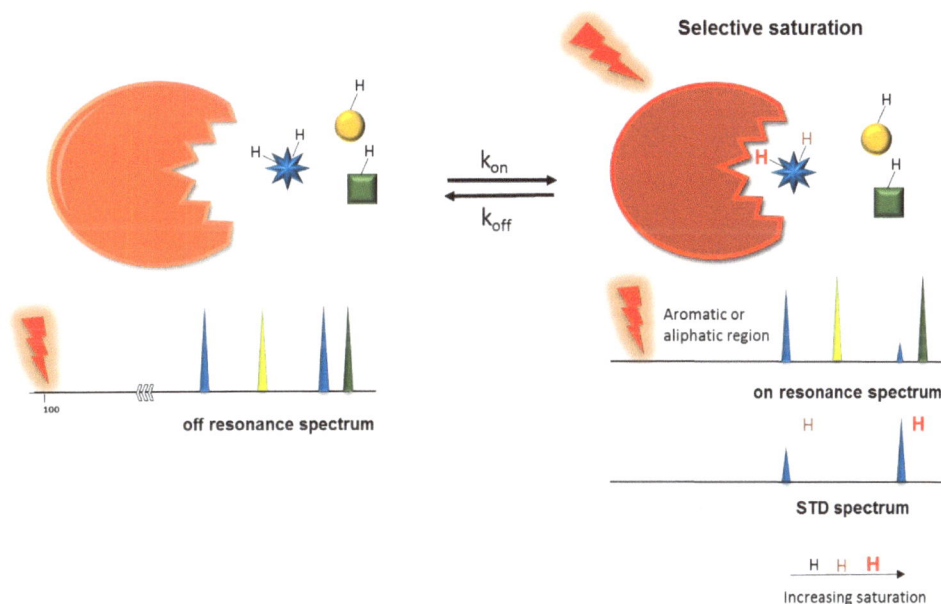

Fig. (15). Schematic representation of the STD-NMR experiment. Resonances of the protein are selectively saturated (aromatic or aliphatic region) and rapidly transferred by spin diffusion throughout the protein. Then, the intermolecular NOE is only transferred to those ligands, with suitable shape and size that interact with the protein. The transference of saturation from the protein to the ligand is detected after subtraction of the on- to off-resonance spectrum in order to obtain a difference spectrum called STD spectrum. The ligand proton signals will be affected differently, depending on their proximity to the protein protons (as shown with different colour intensity and font sizes in the figure), enabling to extract the so-called epitope mapping.

One of the main requirements of the STD-NMR experiment relies on the on-resonance frequency selection (aromatic or aliphatic region). The protein should be selectively (no protons of the ligand should be close to the on-resonance frequency) and efficiently saturated (the duration of the saturation pulse should be optimized). However, long saturation pulses (more selective) lead to a strong spread of magnetization through all protons of the ligand and no epitope mapping can be defined. Basics aspects in acquisition of STD-NMR in explained in more detailed in note 13. Another issue that can cause errors in the binding epitope determination is when the ligand has protons with substantially different values of T1 longitudinal relaxation [59]. In fact, for protons with long T1 values an overestimation in the STD intensity will happen, while those with shorter T1 will be underestimated. For instance, the aromatic protons have long T1 while methylene protons encode short T1 values, thus especially attention should be taken for ligands containing both functional groups. The problem of T1 can be scrutinized applying the STD initial build-up rate method [60], by performing a series of STD experiments at distinct saturation times instead of measuring just a single STD experiment at a specific saturation time. The slope of the STD build-up curve in function of the saturation time will be proportional to the STD intensity; however will be independent of the intrinsic T1 of each proton. STD is particular suitable for ligands with moderate affinity with dissociation constants (K_D) around 10^{-3} to 10^{-7} M. Strong binders will be not observable in the STD spectrum, however can be indirectly detected using a competitive ligand with lower affinity [61].

Hence, the STD technique enables to identify a binding compound in a mixture, as well as to define which proton are in closer contact with the protein. STD-NMR is presently the most accessible and robust NMR-based method for detection of protein-ligand interactions, especially relevant in pharmaceutical industry as a basic method in drug discovery and development [62]. Characterization and recognition of glycan-based tumor-associated antigens by monoclonal antibodies (mAbs) [63] or rational design of small molecules as potential anti-inflammatory drug candidates [64, 65] are only few examples found in the literature that employed STD-NMR as key technique to elucidate the unique features that govern protein-ligand interactions in a medicinal chemistry context.

The STD technique can also be combined with other two-dimensional experiments such as a 2D-STD-TOCSY [66] or 2D-STD-HSQC/HMQC [67]. These STD schemes are particularly useful if applied to large molecules, since it permits to avoid signals' overlapping by increasing the STD signal dispersion [68]. In addition, if the ligand encodes other NMR-active nuclei than protons, alternative filtering/editing arrangements may also be employed to detect ligand-protein interactions. A relevant example is the case of ^{19}F-containing ligands [69]. STD-NMR has also been applied to monitor ligand-receptor interactions in more complex systems such as serum samples [70] and viruses [71] or even using living cells to study membrane proteins [72, 73]. The quantitative analysis of STD data assisted with molecular modelling protocols has been employed to deduce the binding modes of distinct ligand-receptor complexes [74, 75]. In this approach, the CORCEMA-ST (Complete Relaxation and Conformational Exchange MAtrix-Saturation Transfer) method, which allows to predict the STD signal intensities for a specific ligand-protein geometry complex, has been employed [76, 77]. This combined approach allows us to infer that STD-based epitope of GalNAc, deduced in the presence of macrophage galactose-type lectin (MGL), can only be explained if two binding geometries, A and B are present in solution (Fig. **16**).

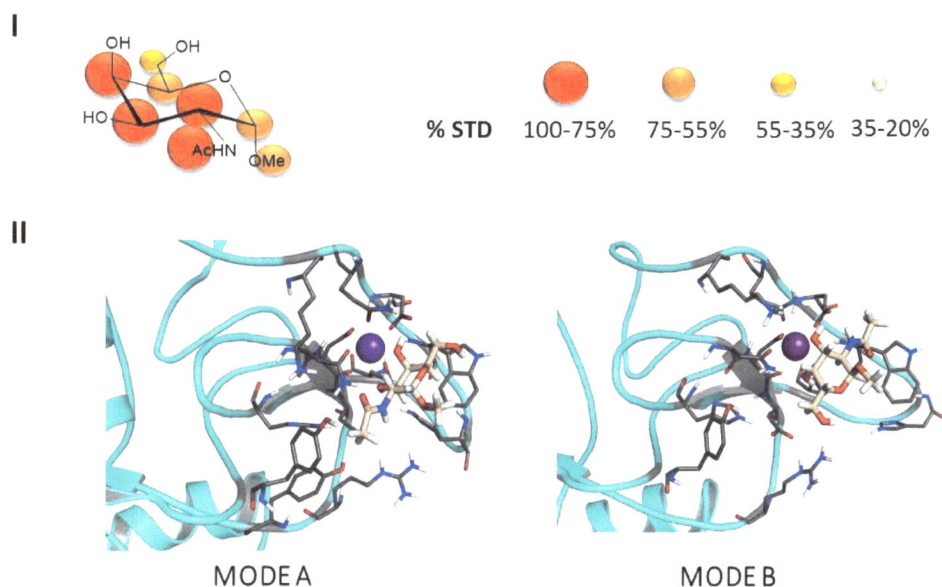

Fig. (**16**). **I.** STD-based epitope mapping obtained for GalNAc with MGL. **II.** Complexes of MGL selected from molecular dynamics calculations of both binding modes A (left panel) and B (right panel) in presence of GalNAc. Both modes, in around 4:1 proportions of modes B and A, enable to explain the STD saturation of the protons of GalNAc.

STD also enables to deduce binding affinities [78]. STD intensities are proportional to the protein–ligand complex concentration in solution, however, for an accurate estimation certain parameters should be optimized, such as saturation time, fraction of bound ligand, affinity, and concentration of receptor (see note

13). Alternatively, STD-NMR competition binding experiments using a reference ligand of known K_D value has been used and allows to determine the inhibitory potency of every ligand competing with the reference compound [53].

2.2.3.2. Diffusion Experiments

In diffusion NMR experiments the interesting observable variable is the self-diffusion coefficient. Self-diffusion coefficients are closely related with the molecular size and shape of the molecule. In this sense, the fact that the diffusion coefficient D of a molecule can be altered by the presence of another molecule if there is an interaction between them is the basis of the application of diffusion NMR to the study of molecular interactions. Diffusion can be investigated by NMR either by analysis of relaxation data [79, 80] or by pulsed-field gradient (PFG) based techniques. While relaxation measurements are sensitive to motions occurring in the psto ns time scale, in PFG experiments motion on the ms to s time scale is measured. Most of the NMR methods developed to measure diffusion are based on PFG NMR. The name DOSY (diffusion ordered spectroscopy) refers to the presentation of the data obtained in PFG NMR measurements, typically in a 2D-Plot (Fig. **17**) where the chemical shift is plotted in one dimension (x-axis) and the diffusion coefficient, D, in the other dimension (y-axis). This presentation allows the identification of signals belonging to one component (or components showing the same diffusion coefficient) which makes DOSY as a "non-invasive chromatography" [81]. Diffusion NMR has been applied to the study of intermolecular interactions both qualitatively, to identify compounds that bind to a specific receptor in NMR screening or in studies related to host-guest chemistry, and quantitatively, in the determination of association constants and complex or aggregate sizes [82]. In fact, when a small molecule binds to a large receptor, as in a ligand binding to a protein, its diffusion coefficient may decrease more than one order of magnitude (Fig. **17**). This means that due to the exchange between the free and the bound state, for some time, the small molecule will encode the diffusion coefficient of the large receptor. The influence of exchange in the DOSY spectra of a two-site system has been considered and analysed in detail in previous reports [83, 84].

The identification of ligands from mixtures, where the diffusion coefficient of a small molecule is altered upon binding to a large receptor, has also been called "Affinity NMR". In the fast exchange limit the observed diffusion coefficient (D_{obs}) is the average of the diffusion coefficient of the molecule in the free and bound states, D_{free} and D_{bound}, respectively, weighted by the respective molar fractions of the species, f_{free} and f_{bound} as depicted in Eq.(7).

$$D_{obs} = f_{free}D_{free} + f_{bound}D_{bound} \qquad (7)$$

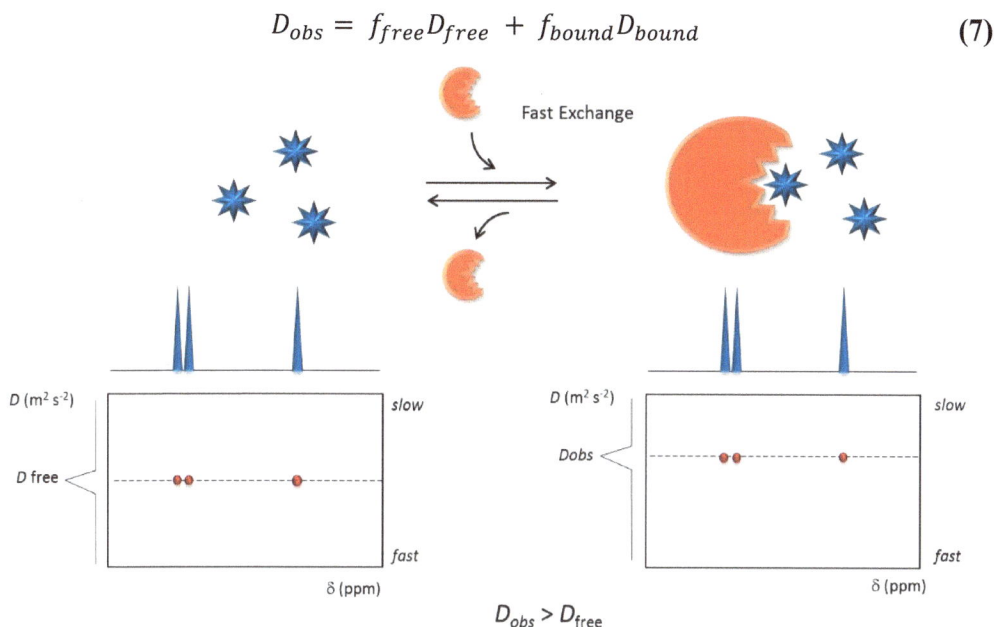

Fig. (17). Schematic representation of the DOSY experiment. Upon binding the diffusion coefficient of the ligand (D_{obs}) will become much larger than in the free state (D_{free}).

Experimentally, DOSY spectrum of the ligand in free state and in the presence of the receptor, are carried out and compared. Due to the linear dependence of D_{obs} with the fraction of bound state low protein-ligand molar ratios are employed. Fig. (**17**) compares the diffusion of a ligand in the absence and presence of the macromolecule. From analysis of DOSY results, the existence of an interaction between the ligand and the protein is clear, since there was a change in the diffusion coefficient of the ligand. DOSY experiments were carried out to discriminate binding of two sugars (cellohexaose and laminarihexaose) in the presence and in the absence of a carbohydrate binding protein (CBM11) [85].

The overlapping of both ligand and receptor signals may cause problems in the calculation of the ligand diffusion coefficient. Therefore, as in the case of STD for complex mixtures a combination of DOSY and TOCSY experiments (DECODES method) can be very useful to avoid signal overlapping issues [86, 87]. Other nuclei can be monitored to overcome proton signal overlapping. For example, by monitoring the ^{19}F spectra different ligands from a mixture of fluorinated molecules in the presence of bovine pancreatic trypsin were discovered [88].

The determination of association constants using DOSY type experiments is also possible and requires quantitative estimation of the diffusion coefficients with precision and accuracy [89]. Comprehensive reviews on NMR diffusion experiments can be found in the literature [90]. Basic aspects concerning to diffusion experiments in protein-ligand binding context are given in note 14.

2.3. Studying Protein Shape and Protein-ligand Interactions by SAXS Methods

Small angle X-ray scattering (SAXS), multi-angle light scattering (MALS) and dynamic light scattering (DLS), are popular techniques among biochemists to characterize macromolecules, providing useful information on oligomerization state, protein aggregation, protein-protein interactions and overall structure (for reviews see [91 - 94]). The techniques rely on the scattering of electromagnetic radiation by the electrons of the molecular matter under analysis. The scattering phenomenon occurs through elastic collisions, where the total kinetic energy of both, the beam and the electrons of the matter, is the same before and after the encounter. In SAXS, a monochromatic beam of X-rays hits the particles in solution and is scattered in all directions, upon constructive and destructive interferences. As described in the previous section, in mono-crystals, diffraction peaks are measured and used to obtain electron density maps and high-resolution structures. However, in SAXS the molecules are not ordered, moving freely in solution. In this case, even though no diffraction peaks are observed, a small deflection of light ($2\theta = 0.1\text{-}10°$, rationale for the name "small-angle") can be collected, forming a scattering pattern. This pattern can be used to derive interatomic distances, which can ultimately provide high-precision information on size and shape of the particles (Fig. **18**).

2.3.1. Data Collection

In a typical SAXS experiment, most of the X-ray beam (produced in an in-house diffractometer or a synchrotron facility) passes through the solution without interacting with particles, while a small part of the beam hits the electrons of the atoms. The electrons act as sources of secondary waves, which arrive at a 2-D X-ray detector, positioned opposite and in-line with the source. The scattering pattern shows the variation of the scattering intensity (I) with the momentum transfer s (also denominated q). s is the difference between the momentum of the X-ray wave before and after the collision with the particles in solution and is determined as $s = (4\pi\sin\theta)/\lambda$ (where 2θ is the angle between the incident and the scattered beam).

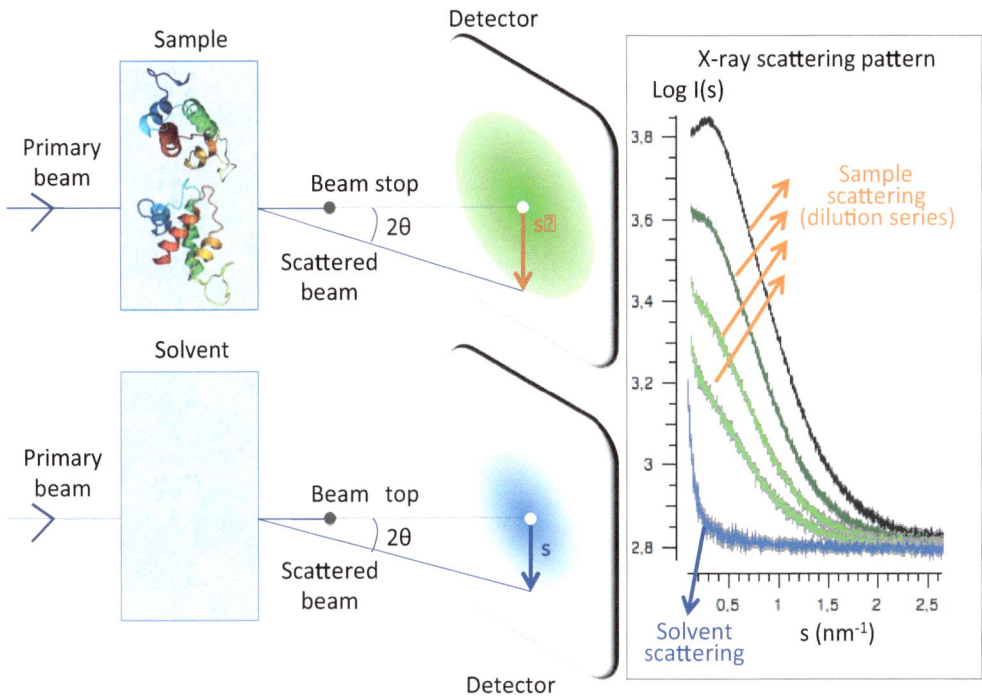

Fig. (18). Overall scheme of a SAXS experiment and the scattering pattern of the sample, at different concentrations, and of the pure solvent. In the scattering pattern, it is possible to observe that the scattering intensities (or log I(s)) is higher in the highest particle concentrations.

The scattering pattern is isotropic and the scattering intensity, which is proportional to the concentration of non-interacting molecules in a monodisperse solution, can be averaged over all orientations Ω. It is usually presented as a one-dimensional curve of radially averaged I as a function of s as shown in Eq. (8).

$$I_{(s)} = <I_{(s)}>_{\Omega} = <A_{(s)}A^*_{(s)}>_{\Omega} \qquad (8)$$

A_s, the scattering amplitude, is the Fourier transformation of the scattering length density.

Biological molecules such as proteins, lipids, carbohydrates or nucleic acids are very poor scatterers when compared to other particles with heavy atoms, as metal alloys or nanostructures. Nevertheless, structural information can be derived from their scattering pattern after subtracting the scattering contribution of the solvent (usually aqueous buffers). In this regard, X-ray exposure and data collection are usually set for pure solvent and for a series of dilutions of the macromolecule under study. The mean difference between the particle and the solvent scattering is termed "contrast" (see Note 8).

Interpretation of a scattering pattern relies on the assumption that the collected data arises from a monodisperse solution of identical particles (see Notes 9 and 10). Biophysical techniques should be applied to investigate the sample quality, which will dictate the degree of confidence in the structural models derived from the SAXS data.

2.3.2. Preliminary Sample Characterization: Guinier; Porod, Kratky and Pair Distribution Functions

Data processing of the scattering intensities was first developed by Andre Guinier in the late 30s to derive the radius of gyration (Rg) of a particle [95]. Nowadays automated data processing software are available, facilitating data analysis and allowing for fast preliminary sample characterization (Fig. **19**). Both manual and automated methods rely on a good estimation of the scattering intensity at zero angle (or forward scattering intensity, I(0)), which cannot be experimentally obtained. At zero angles, the scattering intensity is proportional to the solution concentration and to the number of atoms in the particle (molecular mass, MM). A Guinier plot (ln[I(s)] *versus* s^2) shows a linear region at low angles from where I(0) and Rg can be extracted. When this region is not linear, it indicates potential problems with the sample, either attractive or repulsive interactions, or improper background subtraction, which results in a bad estimation of Rg and MM.

In the 80s, Günther Porod was able to derive the hydrated volume (Vp) of a particle from its scattering pattern [96]. In this approach, the particle is not described at an atomistic level (as in crystal diffraction experiments) but a uniform electron density is assumed inside the particle. Since the electron density is not uniform in macromolecules, the Porod's law requires the subtraction of an appropriate constant from the scattering data, which gives reasonable values for molecules with MM> 30 kDa. The Porod's hydrated volume is independent of the concentration and provides a better estimation of the particle MM and the oligomeric state of the sample.

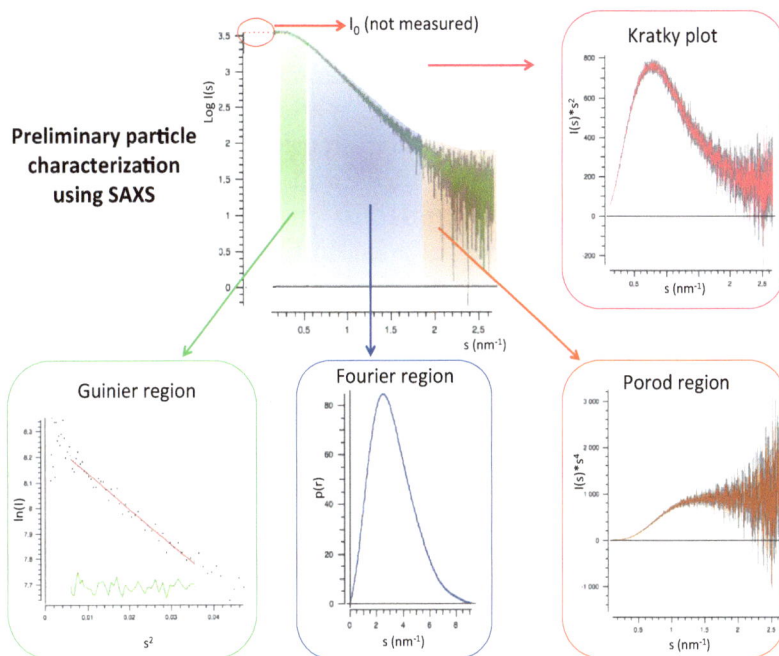

Fig. (19). The scattering pattern can be divided into three regions, each one providing relevant information.

Protein folding and unfolding can be easily evaluated from the scattering data, when displaying $I(s)s^2$ *vs* s, known as the Kratky plot. A globular, properly folded protein, shows a prominent peak at low angles with a plateau at high angles while unfolded proteins show a continuous increase in the scattering in $I(s)s^2$ [97, 98].

Finally, the scattering pattern can provide preliminary information on the shape of the particles by applying a Fourier transform to the scattering intensity. The result is a pair-distance distribution function (P(r)-distribution) showing interatomic vectors (distances between electrons) within the particle [99]. At this stage, analysis of the P(r)-distribution, which is a real space representation where the x-axis is in Angstroms or nanometers, provides useful hints on the overall shape of the particle, allowing to distinguish globular, bell-shaped, rod, multi-domain particles and conformational changes. The pair-distance distribution function also gives a more precise estimation of I(0) and Rg (and MM) regarding the Guinier analysis and similarity between the two reveals internal consistency and robustness of the data.

2.3.3. Deriving a Structural Model

Ab initio algorithms were developed in the late 90s enabling the reconstruction of molecular shapes from scattering data [100, 101]. In an iterative (and automated) form, software is able to describe complex shapes of macromolecule as finite volume elements (dummy atoms and later, dummy residues) resulting in several hypothetical models. Theoretical scattering patterns are calculated from these models and the discrepancy (χ) between the experimental and calculated curves is analyzed for model ranking [102, 103]. The most prominent features will be common to the different models similar to an NMR ensemble [104]. Low-resolution cryo-EM shapes can be used as starting shapes and high-resolution structures can be docked in SAXS models [105]. The combination of X-ray crystallography, cryo-EM and SAXS (and SANS, see Note 11) is usually the best option for characterizing large multi-component complexes, in an integrative approach.

Instead of using *ab initio* methods, theoretical scattering curves can also be calculated from high-resolution models, in case the molecules have been characterized by other techniques (NMR or X-ray crystallography). As previously described, a reasonable SAXS model will generate a theoretical curve that will adequately fit the experimental data.

The final SAXS model can be deposited in a database, similar to what is already established for cryo-EM, NMR and X-ray crystallography. The Small Angle Scattering Biological Data Bank (SASBDB, https://www.sasbdb.org [106]) allows investigators to locate and access experimental scattering data, relevant experimental conditions, sample details, instrument characteristics and the final curated models.

2.4. Structural Elucidation Using Cryo-Electron Microscopy

Cryo-electron microscopy includes different techniques: Electron crystallography, Cryo-electron tomography (Cryo-ET) and single-particle Cryo-EM. While electron crystallography has limited application since ordered entities are required (2D crystals or helical assemblies), Cryo-ET is being widely used and allows imaging structures at a cellular and sub-cellular levels. This kind of tomography is emerging as a powerful tool for imaging all cellular components providing new exciting ways to determine the location of individual biological complexes in cells (see review by Lučić, *et al.*, 2013 on Cryo-ET [107]).

Single particle Cryo-EM, very often abbreviated to Cryo-EM (or SPEM) is the one with the greatest potential as a method of determining the 3D structure of biological macromolecules to (almost) atomic resolution. In the past decade, it became an increasingly powerful technique for structure elucidation of proteins but it provided relatively low-resolution data. It has been mostly applied to the study of large macromolecular complexes that are too large and/or too flexible to be solved by X-ray crystallography or NMR. However, in recent years major advances and improvements in the associated methods and technologies have turned Cryo-EM into a fast-growing area of Structural Biology. It is already considered a high-resolution structure-determination tool [108].

In Cryo-EM a highly homogenous sample of the protein solution is required and after the specimen is plunge-frozen in liquid nitrogen or helium, data from a large number of 2D projection images are obtained (see flow chart in Fig. **20a**). These correspond to identical copies of the protein complex in different orientations that are then combined to generate a 3D reconstruction of the molecular structure. The mathematical process involved is complex but multiple algorithms and computer programs are available.

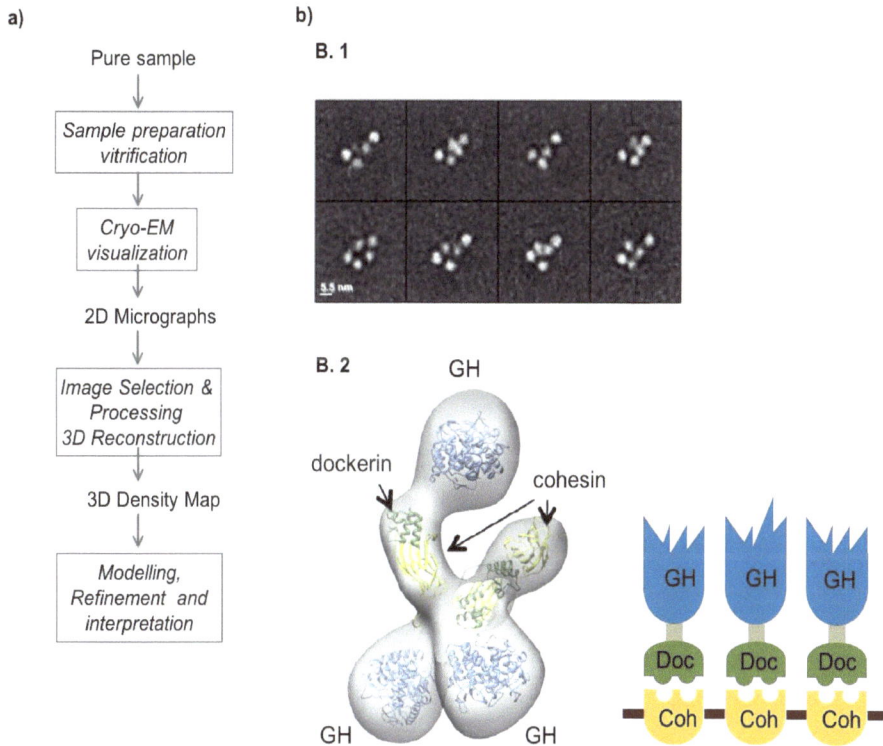

Fig. (20). a) Flow chart of common procedures in single particle reconstruction Cryo-EM (adapted from [139]) b) Cryo-EM structure of a partial cellulosome of *Clostridium thermocellum* to 35Å resolution (single-particle Cryo-EM). Cellulosomes are megaDalton modular assemblies of cellulolytic enzymes and non-catalytic modules excreted by microorganisms that efficiently degrade the plant cell wall polysaccharides. B.1 Representative reference-free 2D averages for the partial cellulosome. B. 2 Cryo-EM envelope showing the fitted atomic structure of cohesin (Coh) and dockerin (Doc) modules, and the glycoside hydrolase (GH) catalytic domain (X-ray structures PDB ID 2CCL (coh-doc) and 1CEM (GH8)) (adapted from [140]).

Recent reviews in the different techniques [109, 110] as well the analysis of the data deposited at the Electron Microscopy Data Bank (EMDB) (http://www.emdatabank.org/) allow to have an excellent overview of the greatest achievements, challenges and potential future applications of Cryo-EM techniques for the determination of 3D structures of biological macromolecules.

To date, X-ray crystallography is the method that accounts for 90% of all structures deposited in the PDB and NMR accounts for 9.4%. Both techniques have their own limitations such as in the analysis of dynamic protein assemblies, difficulty to crystallize or too large in size and in these cases cryo-EM can be particularly well suited. Even when very low-resolution maps are obtained by EM, these can provide important information on the molecular structure and be used in combination with NMR or X-ray structures (see example in Fig. **20b**) [111, 112].

2.5. Notes

2.5.1. Growing and Preserving a Protein Crystal

There are four main methods to grow single crystals from purified proteins, very much dependent on protein availability and the preference of the user: vapor diffusion, micro batch, dialysis and free interface diffusion. In the most widely used vapor diffusion method (Fig. **21**), a drop containing the protein sample and a precipitant solution is set to equilibrate against a reservoir solution containing only the precipitant solution (at a higher concentration than that in the drop). Drops are usually set up with volumes spanning from nanoliters (using automated equipment) to microliters.

In micro batch crystallization, controlled evaporation of water is guaranteed by placing the protein/precipitant mixture under mineral oil. A different approach is present in crystallization by dialysis, where the flow of ions (but not proteins or other polymers) through a dialysis membrane causes a slow super saturation of the protein solution, slow enough to generate single crystals [113]. Capillaries of narrow diameter are used in free interface diffusion to slowly mix a protein solution and a precipitant solution until equilibrium, creating a gradient of concentrations along the capillary. Crystals will appear in the adequate protein: precipitant concentrations' ratio [114].

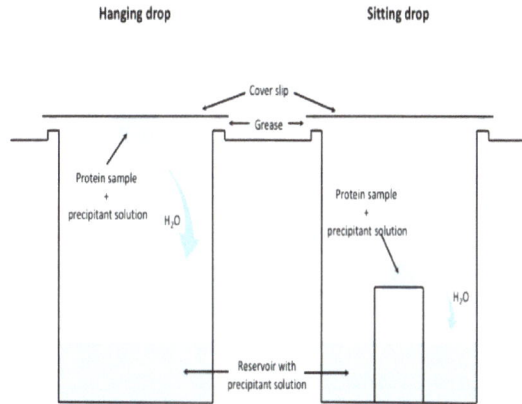

Fig. (21). Two techniques are comprised in the vapor diffusion method: the hanging drop and the sitting drop, as schematized. In both, the protein solution is slowly concentrated in the presence of a precipitant agent and in a sealed reservoir. The protein and precipitant are mixed in the drop, while only the precipitant is in the reservoir at a slightly higher concentration. Equilibration will force water vapor out of the drop, slowly increasing the protein's concentration and, ideally, generating single crystals.

Besides the choice of setup, most important is the rationale behind how to change the individual parameters that can trigger nucleation and influence crystal growth. So far, the phenomenon and its theory are not completely understood, and crystal optimization is still done by changing parameters in a trial-and-error process, but the phase diagram of protein concentration versus precipitant concentration can help "visualizing" what is happening during this process (Fig. **22**) [115].

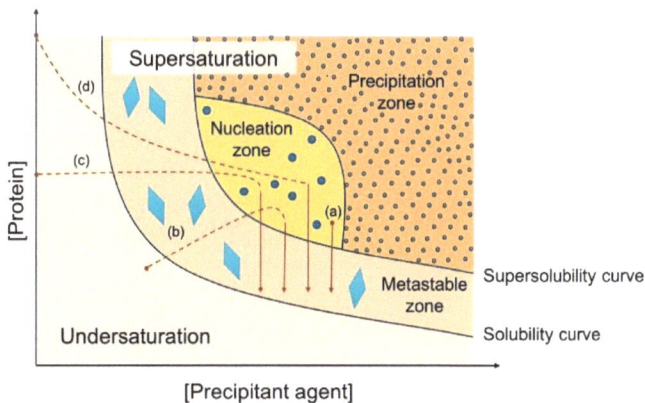

Fig. (22). Schematic illustration of a phase diagram for crystal growth. The paths to obtain protein crystals are represented for the four main crystallization methods: (a) microbatch, (b) vapor diffusion, (c) dialysis and (d) free interface diffusion. The solubility curve is defined as the equilibrium between crystals and protein in the solute at a certain concentration. The supersolubility curve separates the condition where nucleation or precipitation spontaneously occur from the condition where the solution remains clear. Adapted from [141].

After reaching the adequate size, crystals are usually ready for X-ray diffraction experiments. These can be performed at room temperature (around 290K) or under a gas stream of cooled nitrogen (around 100K). Cryo-cooling is particularly important at synchrotron sources for it greatly increases the crystal's resistance to X-ray photons, more specifically to free radical species that generate inside crystals during exposure to X-rays [116]. Crystals can be grown in the presence of a cryoprotectant agent, which will prevent ice formation and subsequent crystal damage and loss of diffraction capabilities, or preformed crystals can be transferred and incubated in a solution containing the cryoprotectant (usually for very short periods prior to plunging in liquid nitrogen or transfer to a cooled nitrogen gas stream) [117]. Common cryoprotectants are glycerol, 2,3-butanediol, ethylene glycol, methyl-pentanediol, different molecular weight polyethylene glycol, paratone oil, sucrose and xylitol [118].

2.5.2. Bragg's Law

Bragg's Law was derived in 1913 by William Henry Bragg and his son William Lawrence Bragg, and shows how the angle of the incident radiation, θ, the distance between planes in a crystal, d, and the wavelength of the incident radiation, λ, are related as shown in Eq. (9):

$$n \times \lambda = 2 \times d \times \sin \theta \qquad (9)$$

The waves of the incident monochromatic radiation are reflected by the parallel planes in the crystal, all equally spaced. When the difference in optical path between the scattered waves is an integer multiple of the wavelength, constructive interference occurs and a diffraction spot is recorded by the detector.

2.5.3. Seleno-methionine Derivatives for Structure Solution

Seleno-amino-acids are naturally occurring residues that have a selenium atom replacing the sulfur atom. Selenium is twice as heavy as sulfur and much heavier than the common elements of the polypeptide chain. Seleno-methionine can be easily incorporated in a protein that is being over-expressed during the cell expression phase. In the cell culture medium, all amino-acids are provided but methionine, which is replaced by seleno-methionine. In many cases, expression conditions need to be tuned for the Se-protein derivative but often enough quantity is available for crystallization. The phase information can be derived upon a SAD experiment at the Se edge (0.9795 Å).

2.5.4. Patterson Maps and Molecular Replacement

Patterson maps are calculated using the electron density equation (Eq. 2), with the exception that the phases are considered to be 0. Eq (2) is simplified to Eq. 10:

$$P(x, y, z) = \frac{1}{V} \sum_{hkl} |F_{hkl}|^2 \cdot e^{-2\pi i(hu+kv+lw)} \tag{10}$$

where u, v, w are the positions of the peaks in the map, corresponding to the interatomic distance vectors [119]. The peaks' height is weighted by the product of the number of electrons in the atoms. A Patterson map of N atoms will have $N(N-1)$ peaks, excluding the central peak (that includes all the self-vectors). Like the diffraction pattern, the Patterson map is centrosymmetric and has the dimensions of the crystal's unit cell. It can be directly calculated from the intensities measured in the diffraction experiment ($I_{hkl}=|F_{hkl}^2|$). The Patterson method is used either to locate heavy atoms in Single or Multiple Isomorphous Replacement methods [120] or to find the unit cell position of a search model in Molecular Replacement methods [121]. The latter are very common when trying to solve the structure of a newly crystallized protein-ligand complex. In such cases, very often the protein's atom coordinates are already known, and this structure is used (and referred to) as the search model. The structure factors calculated from this search model are used to calculate the Patterson map of the model, while the intensities are used to calculate the Patterson map of the new crystal's unit cell. The two Patterson maps are superposed for comparison to find a match (if the known and unknown structures are similar, their Patterson maps should be similar also). Computer software (*e.g.* PHASER [122]) do this in two steps: 1. rotation of the superposed Patterson maps to find an orientation matrix, and 2. translation of the superposed Patterson maps to find a translation vector. The correct positioning of the search model in the new unit cell provides the phase estimates that solve the Phase Problem and enable calculation of the electron density map.

2.5.5. Electron Density Maps

The model building and refinement process is usually guided by the calculation and analysis of two main types of electron density maps (but others exist!): the Fourier simple difference map (usually referred to as F_{obs}-F_{calc} or F_oF_c map) and the double difference map ($2F_{obs}$-F_{calc} or F_o-F_c map), represented with the proper contours. Thorough observation of these maps will help the crystallographer decide which are the appropriate changes to make to the model (used to calculate structure factors: the F_{calc}) bringing it closer to the information given by the diffraction data (the F_{obs}). The map F_o-F_c will reveal positive density peaks where features should be added to the model (supported by the diffraction results) and negative density peaks where features should be removed (because it contradicts the experimental data). The double difference map2 F $_o$-F_c, which fulfills the polypeptide chain and the conserved solvent molecules, is used to minimize the influence (bias) introduced by the calculated phases from the model. Maps representation and observation, as well as model fitting, are done using graphic software interfaces (*e.g.* COOT [21]).

2.5.6. The Temperature Factor (B Factor)

The temperature factor or *B* factor is a thermal vibration parameter that describes the degree of dynamic oscillation of an atom around a defined position. The thermal vibration parameter B_j for atom *j* is a local measure of the relative mobility of different regions of the molecule: B=8∏Uj2=79Uj2; Uj: mean deviation of atom *j* in relation to its average position.

For example: $B_j \approx 80\text{Å}^2 \rightarrow U_j \approx 1\text{Å}; B_j \approx 20\text{Å}^2 \rightarrow U_j \approx 0.5\text{Å}$. The *B* factor factor can also indicate errors in the model building process since, for example, wrongly placed atoms will exhibit higher *B* factors, when compared with neighboring atoms.

2.5.7. The Ramachandran Plot

The Ramachandran plot is a two-dimensional diagram that represents the conformation of all amino acid side chains of a certain protein. The conformation is defined by two main chain dihedral angles (or torsion angles) (ϕ and ψ) for each amino acid residue. The main chain torsion angle phi (ϕ) is defined around the N-Cα bond of a residue (defined by C-N-Cα-C) while the angle psi (ψ) is around the bond Cα-C of a residue (N-Cα-C-N). The Ramachandran plot represents pairs of ϕ- ψ angles in an energy contour plot and defines allowed and forbidden areas of values for both torsion angles. This has to do with the non-repulsive or repulsive van der Waals interactions that may limit the torsion angles in some regions. For very well refined structures, the ϕ, ψ angles of essentially all residues (>95%) must be localized in energetically favourable regions, with the exception of glycine, since it does not have a side chain at Cα.

2.5.8. Technical Aspects of NMR Hardware and Software

^{1}H, ^{13}C, ^{19}F, ^{31}P and ^{15}N are the most common active nuclei used to monitor ligand-protein interactions. Nowadays most NMR laboratories are equipped with cryogenic "inverse" triple resonance proton detection probe heads allowing to improve significantly the NMR experiment sensitivity for ^{1}H, ^{13}C and ^{15}N-experiments. In the case of ^{19}F or ^{31}P nuclei normally different probe heads are required. Gradient units also make part of most NMR spectrometers. Pulse-field gradients (PFG) based methods permit to reduce substantially the duration of NMR experiments. Software for NMR acquisition and basic processing is furnished by the spectrometer manufacturers. Noteworthy, several software packages for analysis of NMR data are available such as, MestreC [123], CARA [124], CCPN [125], SPARKY [126], among others.

2.5.9. NMR Sample

Ligand-protein studies are carried out in solutions that mimics the biological environment thus buffer solutions in H_2O/D_2O 90:10 or D_2O are the best choice. Mostly only non-exchangeable H-C protons of small ligands are monitored in ligand-based NMR methods thus D_2O buffer solutions are mainly used. However, in the case of receptor-detected methods H_2O/D_2O 90:10 mixtures are required to enable the detection of the exchangeable amide protons. Distinct unprotonated (*i.e.,* PBS) or perdeuterated (Tris-d$_{11}$) compounds are commonly used as buffers. A pH around 6-7.5 is commonly used. NMR samples must be transparent without precipitates to allow a good resolution. High quality NMR tubes should be employed, if necessary Shigemi tubes should be used. Depending of the ligand/protein amounts, 3 or 5 mm diameter NMR tubes can be used, with typical 200 and 500 μL of total volume. A standard NMR reference should be present at low concentration (μM). For ^{1}H-NMR spectra in water, the most common standard references are 4,4-dimethyl-4-silapentane-1-sulfonate (DSS) and trimethylsilyl propionate (TSP) (δ = 0.00 ppm). Especially in samples containing H_2O/D_2O 90:10 but also in those cases where D_2O is the main component in the sample, the water signal is very intense compared to the intensity of ligand/protein resonances therefore, efficient solvent suppression is usually required when recording NMR spectra [127 - 129]. The excitation sculpting method, which makes use of pulsed field gradients, avoids large baseline distortions. Distinct shaped pulses (for instance Sinc1.1000 or Squa100.1000) are used for efficient solvent suppression.

2.5.10. General Aspects on Acquisition and Processing of NMR Spectra

After inserting the NMR tube in the appropriate spinner, both are inserted into the spectrometer probe head by an air flow system. Then, to hold the magnetic field as stable as possible the deuterium signal of the solvent (H_2O/D_2O or D_2O) is used in a routine called lock. Further sample tuning and matching should be performed. This procedure consists in adjusting the probe coils to the radiofrequency (RF) of the nucleus or nuclei of interest. Then sample shimming should be done for the fine adjustment of the magnetic field in order to obtain the best possible field homogeneity in the sample active volume. Appropriate calibration of the pulses (durations and potencies) is crucial to optimize the NMR experiments. Therefore, always load a standard proton pulse sequence (ZG) in order to determine the duration of the 90° high power pulse (μs range). The power level for the 90° hard pulse depend on each spectrometer and is calibrated by the manufacturer. Most of the modern spectrometers allow to do all these steps in an automatic or semi-automatic way. Then select the frequency offset, normally placed at the centre of the spectral window. When water suppression is required, the frequency offset has to be the frequency of the water signal. The number of scans (NS) will depend on sample concentration, as well as, on the phase cycling necessary by the specific pulse sequence. To allow the complete relaxation of the spin magnetization between each scan, a value between 1 to 5 s for the relaxation delay (D1) is critical. The resolution of the experiment is determined by the spectral width (SW) and the number of points (TD) and depends on sample complexity. SW will define the highest frequency to be recognized. The receiver gain depends on the sample and is defined accordingly to the input limits of the analogy-digital-converter (ADC) of the spectrometer.

The excitation RF pulse excites all frequencies within a SW. In the receiver coil the detected signal is a combination of all those frequencies resulting in the interferogram - Free induction decay (FID). Fourier transformation of the FID provides the NMR spectrum. The digitisation procedure translates the electrical NMR signal into a binary number proportional to the magnitude of the signal. The acquisition time (AQ) is defined by the digitisation rate (which is dictated by the SW and defines the sampling dwell time - DW) and on how many FID data points are sampled in total (TD). The frequency between adjacent data points in the spectrum is the digital resolution (DR) in Hz/point. It is possible to improve the sensitivity and resolution by applying distinct window functions before Fourier transformation. The phase correction should be performed by eliminating the zero- and first-order phase errors.

2.5.11. Basics on ^{15}N-^1H-HSQC Titrations

Isotope labelling (^{13}C, ^{15}N) of the receptor is imperative for prior specific resonance assignment of the protein. Selective ^{15}N isotopic labelling of the protein is sufficient to record ^{15}N-^1H-HSQC titrations. For the titration experiments, a series of ^{15}N-^1H-HSQC spectra are acquired in which the concentration of protein is maintained constant and the concentration of ligand varied typically from 0 to 5 equivalents. Usually 20 – 300μM of the receptor protein is used. The ligand/protein molar ratio will be dependent of the strength of the ligand-protein interactions and the chemical exchange regime under study. In contrast, to the

NOE-based methods (NOE α r^{-6}) the chemical shift perturbation in the ^{15}N-^1H-HSQC spectrum linearly depends of the complex formation thus small ligand/protein molar ratios are employed. Standard ^{15}N-^1H-HSQC pulse sequences are available in the spectrometers [130 - 133]. The proton central frequency is set on the solvent signal (water) and for nitrogen set on the centre of the amide region.

2.5.12. Basics on Transferred NOE

For suitable analysis of trNOE data, two NOESY experiments of two distinct samples, with and without the receptor, should be recorded. The main difference will be the mixing time of each NOESY experiment that should be adjusted accordingly. In the presence of the protein receptor small mixing times should be used to avoid spin-diffusion, typically around 50-100ms. The small ligands usually need 500ms to 1s of mixing time. Depending on the affinity and kinetic parameters trNOE experiments are recorded in an excess of ligand, usually from 1:5 to 1:50 of protein-ligand molar ratios. Usually 5–100 μM of receptor protein is used. For quantitative analysis, it is essential to record different experiments with variations in the NOE mixing times and using different ligand/protein molar ratios. Standard NOESY pulse sequence with water suppression are currently available in the spectrometers [128]. For homonuclear^1H-^1H-NOESY spectra the central frequency is set on the water signal.

2.5.13. Basics on STD-NMR Experiments

Experimentally in the STD-NMR two ^1H-NMR experiments are recorded: the on-resonance and off-resonance spectra. The selective saturation pulse on the on-resonance spectrum should be set at frequencies at which only protons signals of the receptor are present (0 to −1 ppm or around 7 ppm must be found) and with no small ligand resonances appearing at least within 1-2 ppm. The off-resonance spectrum is a reference experiment achieved by saturating a region of the spectrum that does not contain any signal (100ppm). Subtraction of the on- to the off-resonance spectrum yields a spectrum where only those resonances that have received saturation from the protein, namely receptor protons and protons of those ligands that bind to the protein receptor, will be visible. The duration of the saturation pulse should be optimized and distinct STD-NMR experiments should be recorded from 0.5 to 3 sec of saturation time. The saturation pulse is a shaped pulse and should be calibrated. Normally Eburp2.1000 or Gauss1.1000 are selected. The common length of saturation pulse is 50ms with a spacing of 1 ms between repetitions. The potency of the saturation pulse can be determined in function of the previous 90° hard pulse length. To eliminate the background resonance signals of the protein a T1ρ filter of (15-30 ms spin-lock pulse) can be used. Even using deuterated buffer solutions water suppression is sometimes applied. Comparing to the trNOESY, STD-NMR experiment requires smaller amounts of the large receptor (ca. 5-20 µM, MW > 15kDa) and 1:10 to 1:100 protein-ligand molar ratios are recommended. STD-NMR pulse sequences with and without water suppression, with and without spin-lock filter are available in the spectrometers [57]. The STD can be analysed using the amplification factors according to the Eq. (11):

$$A_{STD} = \frac{(I_0 - I_{sat})}{I_0} \times molar\ ratio = \frac{I_{STD}}{I_0} \times molar\ ratio \qquad (11)$$

Where I_0, I_{sat} and I_{STD} are the intensities of the off resonance, on resonance and difference spectrum, respectively. A_{STD} or simply the I_{STD}/I_0 ratio allows to determine the epitope mapping [66]. For that the STD signal with the highest intensity is set to 100% and the STD percentages intensities of the other protons are calculated accordingly. K_D estimation is also possible by recording STD-NMR experiments at distinct concentrations of the ligand. A related equation with the well-known Michaelis-Menten equation can be written for A_{STD} Eq (12):

$$A_{STD} = \frac{\alpha_{STD} \times [L]}{K_D + [L]} \qquad (12)$$

where α_{STD} is the maximum amplification factor and [L] is the ligand concentration. A_{STD} will raise until reached the α_{STD} that happens when [L] >>K_D and the receptor binding site is totally saturated. In K_D determination a more accurate examination of STD-NMR data is based on the STD initial growth rates analysis [78]. A protocol for acquisition, processing and data analysis of STD (including K_D estimation) applied for students in a laboratory and classroom context is clear presented in the literature [134].

2.5.14. Basics on Diffusion Experiments

Diffusion binding experiments are based on the fact that the diffusion coefficient *D* of a small ligand is larger upon binding to the receptor. Therefore, *D* values of the ligand should be extract in free and in presence of the large receptor. Herein, the DOSY experiment corresponds to the acquisition of a series of 1D spectra raising the gradient strength with a fixed diffusion time. Then it is possible to correlate the intensity of the signal, signal attenuation, with the increasing of gradient strength and construct the DOSY spectrum. Nevertheless, it is necessary to optimize the gradient strength, the diffusion time and gradient length for the sample. Easily by running several 2D and changing the diffusion time and gradient strength is enough to find the optimal signal decay profile. The gradient strength determines the signal attenuation while the diffusion time only affects the exponential decay function linearly. For each 2D-DOSY experiment it is also required to define the initial (5%) and final values (95%) of the gradient ramp, the number of steps and the type of the ramp (linear). Diffusion coefficients are linearly dependent of the complex formation thus small ligand/protein molar ratios are employed. In addition, DOSY experiments are much less sensitive than NOE-based methods therefore normally more amounts of sample are necessary. Also, may be imperative to increase the NS to be able to detect the signal attenuation. Usually D_2O buffer solutions are employed to prepare the sample and an internal reference such as TSP should be added to account for viscosity changes. Standard DOSY pulse programs are currently available in the spectrometers [135].

2.5.15. SAXS Buffer Subtraction

Buffer subtraction is a delicate issue in data analysis and inaccuracies can interfere with modeling calculations. In order to assure a good buffer matching, dialysis of the macromolecule with the appropriate solvent is usually the most efficient method. The presence of free ligands should also be taken into account when studying interactions and conformational changes of macromolecules. In this sense, SAXS data is collected in sequential runs for the dialysis buffer and for the macromolecules in the same buffer solution. A poor buffer subtraction can be detected in the pair-distribution function releasing the requirement for P(r)-distribution=0 at r=0. In this case, positive or negative P(r)-distribution at r=0 indicates under- or over subtraction of the buffer, respectively.

2.5.16. Sample Quality for Successful Bio SAXS Experiment

High sample quality is a requirement for acquiring interpretable SAXS data. Protein/DNA impurities contribute for the scattering curves, misleading structure determination (especially if contaminants have higher MM than the macromolecule under study since the scattering intensity is proportional to the square of the MM). Similarly, protein aggregation should be avoided and DLS is widely used to monitor monodispersity and sample oligomerization.

2.5.17. Radiation Damage

Radiation damage can also occur in SAXS experiments since the X-rays will promote bond breakage and free radical formation, which can result in aggregation. The presence of free-radical absorbers as ascorbate or DTT in the buffer is a good strategy to avoid radiation damage. When using synchrotron sources, it is a common procedure to collect several exposures and discard the ones where sample aggregation is observed. Also, a constant flow of the sample through the X-ray beam decreases the exposure time of each individual molecule and decreases the chances of radiation damage.

2.5.18. Small Angle Neutron Scattering (SANS)

Small Angle Neutron Scattering (SANS) is very often combined with SAXS for deriving overall shapes of complexes. It is achieved with the elastic scattering of neutrons (produced by a small in-house device or a nuclear reactor), upon interaction with the atomic nuclei of the molecules in solution. The scattering of deuterium is significantly different from hydrogen and taking advantage of this difference enables the study of multi-component complexes (protein-protein or protein-nucleic acids), where one of the partners is selectively deuterated. Contrast variations series tune the visible/invisible ratio of the scattering of the molecules in solution [136 - 138].

CONSENT FOR PUBLICATION

Not applicable.

CONFLICT OF INTEREST

The author (editor) declares no conflict of interest, financial or otherwise.

ACKNOWLEDGEMENTS

The authors acknowledge Doctor Benedita Pinheiro and PhD students Raquel Costa, Viviana Correia and Ana Diniz and Raquel Santos for assistance in the preparation of Figures and Dr. Haydyn Mertens for critical reading SAXS on the manuscript. Filipa Marcelo thanks Portuguese Science and Technology Foundation (FCT-MCTES) for the post-doctoral grant SFRH/BPD/110734/2015. The X-ray crystallography and NMR content of this book chapter was written under the scope of project RECI/BBB-BEP/0124/2012 (*Modern Structural Biology: Resources for the advancement of in-house X-ray Crystallography*) and RECI/BBB-BQB/0230/2012 (*Nuclear Magnetic Resonance: from Molecular Structure and Dynamics to Protein Function, Cell Physiology, and Metabolomics*) from FCT-MCTES, respectively. The authors also thank the COST action CM1407 (*Challenging organic syntheses inspired by nature: from natural products chemistry to drug discovery*). The Applied Biomolecular Sciences Unit, UCIBIO (Unidade de CiênciasBiomolecularesAplicadas), is financed by national funds from FCT-MCTES (UID/Multi/04378/2013) and co-financed by the

FEDER under the PT2020 Partnership Agreement (POCI-01-0145-FE-ER-007728).

REFERENCES

[1]	http://www.rcsb.org/pdb/statistics/contentGrowthChart.do?content=explMethod-

[2]	M. Pellecchia, I. Bertini, D. Cowburn, C. Dalvit, E. Giralt, W. Jahnke, T.L. James, S.W. Homans, H. Kessler, C. Luchinat, B. Meyer, H. Oschkinat, J. Peng, H. Schwalbe, and G. Siegal, "Perspectives on NMR in drug discovery: a technique comes of age", *Nat. Rev. Drug Discov.,* vol. 7, no. 9, pp. 738-745, 2008.
[http://dx.doi.org/10.1038/nrd2606] [PMID: 19172689]

[3]	C.A. Lepre, J.M. Moore, and J.W. Peng, "Theory and applications of NMR-based screening in pharmaceutical research", *Chem. Rev.,* vol. 104, no. 8, pp. 3641-3676, 2004.
[http://dx.doi.org/10.1021/cr030409h] [PMID: 15303832]

[4]	B. Kellam, P.A. De Bank, and K.M. Shakesheff, "Chemical modification of mammalian cell surfaces", *Chem. Soc. Rev.,* vol. 32, no. 6, pp. 327-337, 2003.
[http://dx.doi.org/10.1039/b211643j] [PMID: 14671788]

[5]	S. Velankar, G. van Ginkel, Y. Alhroub, G.M. Battle, J.M. Berrisford, M.J. Conroy, J.M. Dana, S.P. Gore, A. Gutmanas, P. Haslam, P.M. Hendrickx, I. Lagerstedt, S. Mir, M.A. Fernandez Montecelo, A. Mukhopadhyay, T.J. Oldfield, A. Patwardhan, E. Sanz-García, S. Sen, R.A. Slowley, M.E. Wainwright, M.S. Deshpande, A. Iudin, G. Sahni, J. Salavert Torres, M. Hirshberg, L. Mak, N. Nadzirin, D.R. Armstrong, A.R. Clark, O.S. Smart, P.K. Korir, and G.J. Kleywegt, "PDBe: improved accessibility of macromolecular structure data from PDB and EMDB", *Nucleic Acids Res.,* vol. 44, no. D1, pp. D385-D395, 2016.
[http://dx.doi.org/10.1093/nar/gkv1047] [PMID: 26476444]

[6]	A. McPherson, C. Nguyen, and R. Cudney, "The Role of Small Molecule Additives and Chemical Modification in Protein Crystallization", *Cryst. Growth Des.,* vol. 11, no. 5, pp. 1469-1474, 2011.
[http://dx.doi.org/10.1021/cg101308r]

[7]	M.C. Deller, L. Kong, and B. Rupp, "Protein stability: a crystallographer's perspective", *Acta Crystallogr. F Struct. Biol. Commun.,* vol. 72, no. Pt 2, pp. 72-95, 2016.
[http://dx.doi.org/10.1107/S2053230X15024619] [PMID: 26841758]

[8]	S.B. Larson, J.S. Day, R. Cudney, and A. McPherson, "A novel strategy for the crystallization of proteins: X-ray diffraction validation", *Acta Crystallogr. D Biol. Crystallogr.,* vol. 63, no. Pt 3, pp. 310-318, 2007.
[http://dx.doi.org/10.1107/S0907444906053303] [PMID: 17327668]

[9]	F.H. Niesen, H. Berglund, and M. Vedadi, "The use of differential scanning fluorimetry to detect ligand interactions that promote protein stability", *Nat. Protoc.,* vol. 2, no. 9, pp. 2212-2221, 2007.http://eutils.ncbi.nlm.nih.gov/entrez/eutils/elink.fcgi?dbfrom=pubmed&id=17853878&retmode=ref&cmd=prlinks [Internet].
[http://dx.doi.org/10.1038/nprot.2007.321] [PMID: 17853878]

[10]	L. Reinhard, H. Mayerhofer, A. Geerlof, J. Mueller-Dieckmann, and M.S. Weiss, "Optimization of protein buffer cocktails using Thermofluor", *Acta Crystallogr. Sect. F Struct. Biol. Cryst. Commun.,* vol. 69, no. Pt 2, pp. 209-214, 2013.
[http://dx.doi.org/10.1107/S1744309112051858] [PMID: 23385769]

[11]	I.W. McNae, D. Kan, and G. Kontopidis, "Studying protein–ligand interactions using protein crystallography", *Crystallogr. Rev.,* vol. 11, no. 1, pp. 61-71, 2005.http://www.tandfonline.com/doi/full/10.1080/08893110500078639#abstract [Internet].
[http://dx.doi.org/10.1080/08893110500078639]

[12]	A. McPherson, and B. Cudney, "Searching for silver bullets: an alternative strategy for crystallizing

macromolecules", *J. Struct. Biol.,* vol. 156, no. 3, pp. 387-406, 2006.
[http://dx.doi.org/10.1016/j.jsb.2006.09.006] [PMID: 17101277]

[13]   D.G. Waterman, G. Winter, and R.J. Gildea, *Diffraction-geometry refinement in the DIALS framework.,*
2016.http://www.ncbi.nlm.nih.gov/pubmed/27050135www.pubmedcentral.nih.gov/articlerender.fcgi?
artid=PMC4822564scripts.iucr.org/cgi-bin/paper?S2059798316002187

[14]   T.G.G. Battye, L. Kontogiannis, O. Johnson, H.R. Powell, and A.G. Leslie, "iMOSFLM: a new graphical interface for diffraction-image processing with MOSFLM", *Acta Crystallogr. D Biol. Crystallogr.,* vol. 67, no. Pt 4, pp. 271-281, 2011.
[http://dx.doi.org/10.1107/S0907444910048675] [PMID: 21460445]

[15]   G. Winter, "Xia2: An expert system for macromolecular crystallography data reduction", *J. Appl. Cryst.,* vol. 43, no. 1, pp. 186-190, 2010.
[http://dx.doi.org/10.1107/S0021889809045701]

[16]   W. Kabsch, "XDS", *Acta Crystallogr. D Biol. Crystallogr.,* vol. 66, no. Pt 2, pp. 125-132, 2010.
[http://dx.doi.org/10.1107/S0907444909047337] [PMID: 20124692]

[17]   B. Rupp, *Biomolecular Crystallography: Principles, Practice, and Application to Structural Biology.* 1st ed. Garland Science: New York, 2009.

[18]   W.A. Hendrickson, and C.M. Ogata, "Phase determination from multiwavelength anomalous diffraction measurements", *Methods Enzymol.,* vol. 276, pp. 494-523, 1997.
[http://dx.doi.org/10.1016/S0076-6879(97)76074-9] [PMID: 27799111]

[19]   E. de La Fortelle, and G. Bricogne, "Maximum-likelihood heavy-atom parameter refinement for multiple isomorphous replacement and multiwavelength anomalous diffraction methods", *Methods Enzymol.,* vol. 276, pp. 472-494, 1997.
[http://dx.doi.org/10.1016/S0076-6879(97)76073-7] [PMID: 27799110]

[20]   M.D. Winn, C.C. Ballard, K.D. Cowtan, E.J. Dodson, P. Emsley, P.R. Evans, R.M. Keegan, E.B. Krissinel, A.G. Leslie, A. McCoy, S.J. McNicholas, G.N. Murshudov, N.S. Pannu, E.A. Potterton, H.R. Powell, R.J. Read, A. Vagin, and K.S. Wilson, "Overview of the CCP4 suite and current developments", *Acta Crystallogr. D Biol. Crystallogr.,* vol. 67, no. Pt 4, pp. 235-242, 2011.
[http://dx.doi.org/10.1107/S0907444910045749] [PMID: 21460441]

[21]   P. Emsley, B. Lohkamp, W.G. Scott, and K. Cowtan, "Features and development of Coot", *Acta Crystallogr. D Biol. Crystallogr.,* vol. 66, no. Pt 4, pp. 486-501, 2010.
[http://dx.doi.org/10.1107/S0907444910007493] [PMID: 20383002]

[22]   R.A. Engh, and R. Huber, "Accurate bond and angle parameters for X-ray protein structure refinement", *Acta Crystallogr. A,* vol. 47, no. 4, pp. 392-400, 1991.
[http://dx.doi.org/10.1107/S0108767391001071]

[23]   D.E. Tronrud, D.S. Berkholz, and P.A. Karplus, "Using a conformation-dependent stereochemical library improves crystallographic refinement of proteins", *Acta Crystallogr. D Biol. Crystallogr.,* vol. 66, no. Pt 7, pp. 834-842, 2010.
[http://dx.doi.org/10.1107/S0907444910019207] [PMID: 20606264]

[24]   G.J. Kleywegt, and T.A. Jones, "Where freedom is given, liberties are taken", *Structure,* vol. 3, no. 6, pp. 535-540, 1995.
[http://dx.doi.org/10.1016/S0969-2126(01)00187-3] [PMID: 8590014]

[25]   "R factor - Online Dictionary of Crystallography." [Online]. Available: http://reference.iucr.org/dictionary/R_factor

[26]   A.T. Brünger, "Free R value: a novel statistical quantity for assessing the accuracy of crystal structures", *Nature,* vol. 355, no. 6359, pp. 472-475, 1992.
[http://dx.doi.org/10.1038/355472a0] [PMID: 18481394]

[27]   V.B. Chen, W.B. Arendall III, J.J. Headd, D.A. Keedy, R.M. Immormino, G.J. Kapral, L.W. Murray,

J.S. Richardson, and D.C. Richardson, "MolProbity: all-atom structure validation for macromolecular crystallography", *Acta Crystallogr. D Biol. Crystallogr.,* vol. 66, no. Pt 1, pp. 12-21, 2010.
[http://dx.doi.org/10.1107/S0907444909042073] [PMID: 20057044]

[28]   P.D. Adams, P.V. Afonine, G. Bunkóczi, V.B. Chen, I.W. Davis, N. Echols, J.J. Headd, L.W. Hung, G.J. Kapral, R.W. Grosse-Kunstleve, A.J. McCoy, N.W. Moriarty, R. Oeffner, R.J. Read, D.C. Richardson, J.S. Richardson, T.C. Terwilliger, and P.H. Zwart, "PHENIX: a comprehensive Python-based system for macromolecular structure solution", *Acta Crystallogr. D Biol. Crystallogr.,* vol. 66, no. Pt 2, pp. 213-221, 2010.
[http://dx.doi.org/10.1107/S0907444909052925] [PMID: 20124702]

[29]   M.P. Williamson, "Using chemical shift perturbation to characterise ligand binding", *Prog. Nucl. Magn. Reson. Spectrosc.,* vol. 73, pp. 1-16, 2013.
[http://dx.doi.org/10.1016/j.pnmrs.2013.02.001] [PMID: 23962882]

[30]   H. S. Atreya, Ed., Isotope Labeling in Biomolecular NMR. Netherlands: Springer, 2012.

[31]   K. Pervushin, R. Riek, G. Wider, and K. Wüthrich, "Attenuated T2 relaxation by mutual cancellation of dipole-dipole coupling and chemical shift anisotropy indicates an avenue to NMR structures of very large biological macromolecules in solution", *Proc. Natl. Acad. Sci. USA,* vol. 94, no. 23, pp. 12366-12371, 1997.
[http://dx.doi.org/10.1073/pnas.94.23.12366] [PMID: 9356455]

[32]   L. Berliner, *Protein NMR: Modern Techniques and Biomedical Applications.* Springer, 2015.
[http://dx.doi.org/10.1007/978-1-4899-7621-5]

[33]   A.K. Downing, Ed., *Protein NMR Techniques - Methods in Molecular Biology.* vol. 27. Humana Press, 2004.
[http://dx.doi.org/10.1385/1592598099]

[34]   G. Lipari, and A. Szabo, "Model-free approach to the interpretation of nuclear magnetic resonance relaxation in macromolecules. 1. Theory and range of validity", *J. Am. Chem. Soc.,* vol. 104, no. 17, pp. 4546-4559, 1982.http://pubs.acs.org/doi/abs/10.1021/ja00381a009 [Internet].
[http://dx.doi.org/10.1021/ja00381a009]

[35]   G. Lipari, and A. Szabo, "Model-free approach to the interpretation of nuclear magnetic resonance relaxation in macromolecules. 2. Analysis of experimental results", *J. Am. Chem. Soc.,* vol. 104, no. 17, pp. 4559-4570, 1982.http://pubs.acs.org/doi/abs/10.1021/ja00381a010 [Internet].
[http://dx.doi.org/10.1021/ja00381a010]

[36]   A.G. Palmer III, "NMR characterization of the dynamics of biomacromolecules", *Chem. Rev.,* vol. 104, no. 8, pp. 3623-3640, 2004.
[http://dx.doi.org/10.1021/cr030413t] [PMID: 15303831]

[37]   A. Viegas, J. Sardinha, F. Freire, D.F. Duarte, A.L. Carvalho, C.M. Fontes, M.J. Romão, A.L. Macedo, and E.J. Cabrita, "Solution structure, dynamics and binding studies of a family 11 carbohydrate-binding module from Clostridium thermocellum (CtCBM11)", *Biochem. J.,* vol. 451, no. 2, pp. 289-300, 2013.http://www.ncbi.nlm.nih.gov/pubmed/23356867 [Internet].
[http://dx.doi.org/10.1042/BJ20120627] [PMID: 23356867]

[38]   W. Jahnke, "Perspectives of biomolecular NMR in drug discovery: the blessing and curse of versatility", *J. Biomol. NMR,* vol. 39, no. 2, pp. 87-90, 2007.
[http://dx.doi.org/10.1007/s10858-007-9183-5] [PMID: 17701274]

[39]   A.L. Carvalho, A. Goyal, and J.A.M. Prates, "The family 11 carbohydrate-binding module of Clostridium thermocellum Lic26A-Cel5E accommodatesβ-1,4- andβ-1,3-1,4-mixed linked glucans at a single binding site", *J. Biol. Chem.,* vol. 279, no. 33, pp. 34785-34793, 2004.
[http://dx.doi.org/10.1074/jbc.M405867200] [PMID: 15192099]

[40]   L. Fielding, "NMR methods for the determination of protein-ligand dissociation constants. Vol. 51", *Prog. Nucl. Magn. Reson. Spectrosc.,* pp. 219-242, 2007.
[http://dx.doi.org/10.1016/j.pnmrs.2007.04.001]

[41]   I. Kucharska, B. Liang, and N. Ursini, *Molecular Interactions of Lipopolysaccharide with an Outer Membrane Protein from Pseudomonas aeruginosa Probed by Solution NMR.,* 2016.http://pubs.acs.org/doi/abs/10.1021/acs.biochem.6b00630
[http://dx.doi.org/10.1021/acs.biochem.6b00630]

[42]   A. Viegas, N.F. Brás, N.M.F.S.A. Cerqueira, P.A. Fernandes, J.A. Prates, C.M. Fontes, M. Bruix, M.J. Romão, A.L. Carvalho, M.J. Ramos, A.L. Macedo, and E.J. Cabrita, "Molecular determinants of ligand specificity in family 11 carbohydrate binding modules: an NMR, X-ray crystallography and computational chemistry approach", *FEBS J.,* vol. 275, no. 10, pp. 2524-2535, 2008.
[http://dx.doi.org/10.1111/j.1742-4658.2008.06401.x] [PMID: 18422658]

[43]   V. García-Aparicio, M. Sollogoub, Y. Blériot, V. Colliou, S. André, J.L. Asensio, F.J. Cañada, H.J. Gabius, P. Sinaÿ, and J. Jiménez-Barbero, "The conformation of the C-glycosyl analogue of N-acety--lactosamine in the free state and bound to a toxic plant agglutinin and human adhesion/growth-regulatory galectin-1", *Carbohydr. Res.,* vol. 342, no. 12-13, pp. 1918-1928, 2007.
[http://dx.doi.org/10.1016/j.carres.2007.02.034] [PMID: 17408600]

[44]   D. Neuhaus, and M. Williamson, *The Nuclear Overhauser Effect in Structural and Conformational Analysis.* VCH New York Weinheim Cambridge, 2000.

[45]   A.A. Bothner-by, and J.H. Noggle, "Time development of nuclear overhauser effects in multispin systems", *J. Am. Chem. Soc.,* vol. 1979, no. 14, pp. 5152-5155, 1979.
[http://dx.doi.org/10.1021/ja00512a006]

[46]   F. Ni, "Recent developments in transferred NOE methods," Prog. Nucl. Magn. Reson. Spectrosc., vol. 26, pp. 517–606, 1994.

[47]   V.L. Bevilacqua, Y. Kim, and J.H. Prestegard, "Conformation of β-methylmelibiose bound to the ricin B-chain as determined from transferred nuclear Overhauser effects", *Biochemistry,* vol. 31, no. 39, pp. 9339-9349, 1992.
[http://dx.doi.org/10.1021/bi00154a003] [PMID: 1390719]

[48]   J. Revuelta, T. Vacas, M. Torrado, F. Corzana, C. Gonzalez, J. Jiménez-Barbero, M. Menendez, A. Bastida, and J.L. Asensio, "NMR-based analysis of aminoglycoside recognition by the resistance enzyme ANT(4′): the pattern of OH/NH3(+) substitution determines the preferred antibiotic binding mode and is critical for drug inactivation", *J. Am. Chem. Soc.,* vol. 130, no. 15, pp. 5086-5103, 2008.
[http://dx.doi.org/10.1021/ja076835s] [PMID: 18366171]

[49]   A. Canales, R. Matesanz, N.M. Gardner, M. Andreu, I. Paterson, J.F. Díaz, and J. Jiménez-Barbero, "The bound conformation of microtubule-stabilizing agents : NMR insights into the bioactive 3D structure of discodermolide and dictyostatin", *Chem. Eur. J.,* vol. 14, pp. 7557-7569, 2008.

[50]   A. Canales, L. Nieto, J. Rodríguez-Salarichs, P.A. Sánchez-Murcia, C. Coderch, A. Cortés-Cabrera, I. Paterson, T. Carlomagno, F. Gago, J.M. Andreu, K.H. Altmann, J. Jiménez-Barbero, and J.F. Díaz, "Molecular recognition of epothilones by microtubules and tubulin dimers revealed by biochemical and NMR approaches", *ACS Chem. Biol.,* vol. 9, no. 4, pp. 1033-1043, 2014.
[http://dx.doi.org/10.1021/cb400673h] [PMID: 24524625]

[51]   F. Marcelo, S. Huecas, L.B. Ruiz-Ávila, F.J. Cañada, A. Perona, A. Poveda, S. Martín-Santamaría, A. Morreale, J. Jiménez-Barbero, and J.M. Andreu, "Interactions of bacterial cell division protein FtsZ with C8-substituted guanine nucleotide inhibitors. A combined NMR, biochemical and molecular modeling perspective", *J. Am. Chem. Soc.,* vol. 135, no. 44, pp. 16418-16428, 2013.
[http://dx.doi.org/10.1021/ja405515r] [PMID: 24079270]

[52]   R.L. Lock, and E.J. Harry, "Cell-division inhibitors: new insights for future antibiotics", *Nat. Rev. Drug Discov.,* vol. 7, no. 4, pp. 324-338, 2008.
[http://dx.doi.org/10.1038/nrd2510] [PMID: 18323848]

[53]   F. Marcelo, C. Dias, A. Martins, P.J. Madeira, T. Jorge, M.H. Florêncio, F.J. Cañada, E.J. Cabrita, J. Jiménez-Barbero, and A.P. Rauter, "Molecular recognition of rosmarinic acid from Salvia sclareoides extracts by acetylcholinesterase: a new binding site detected by NMR spectroscopy", *Chemistry,* vol.

19, no. 21, pp. 6641-6649, 2013.
[http://dx.doi.org/10.1002/chem.201203966] [PMID: 23536497]

[54]   A. Canales, J. Angulo, R. Ojeda, M. Bruix, R. Fayos, R. Lozano, G. Giménez-Gallego, M. Martín-Lomas, P.M. Nieto, and J. Jiménez-Barbero, "Conformational flexibility of a synthetic glycosylaminoglycan bound to a fibroblast growth factor. FGF-1 recognizes both the (1)C(4) and (2)S(O) conformations of a bioactive heparin-like hexasaccharide", *J. Am. Chem. Soc.,* vol. 127, no. 16, pp. 5778-5779, 2005.
[http://dx.doi.org/10.1021/ja043363y] [PMID: 15839662]

[55]   L. Nieto, Á. Canales, G. Giménez-Gallego, P.M. Nieto, and J. Jiménez-Barbero, "Conformational selection of the AGA*IA(M) heparin pentasaccharide when bound to the fibroblast growth factor receptor", *Chemistry,* vol. 17, no. 40, pp. 11204-11209, 2011.
[http://dx.doi.org/10.1002/chem.201101000] [PMID: 21922554]

[56]   V.M. Sánchez-Pedregal, M. Reese, J. Meiler, M.J.J. Blommers, C. Griesinger, and T. Carlomagno, "The INPHARMA method: protein-mediated interligand NOEs for pharmacophore mapping", *Angew. Chem. Int. Ed. Engl.,* vol. 44, no. 27, pp. 4172-4175, 2005.
[http://dx.doi.org/10.1002/anie.200500503] [PMID: 15929149]

[57]   M. Mayer, and B. Meyer, "Characterization of ligand binding by saturation transfer difference NMR spectroscopy", *Angew. Chem. Int. Ed. Engl.,* vol. 38, no. 12, pp. 1784-1788, 1999.
[http://dx.doi.org/10.1002/(SICI)1521-3773(19990614)38:12<1784::AID-ANIE1784>3.0.CO;2-Q] [PMID: 29711196]

[58]   J. Telser, R. Davydov, Y.C. Horng, S.W. Ragsdale, and B.M. Hoffman, "Cryoreduction of methyl-coenzyme M reductase: EPR characterization of forms, MCR(ox1) and MCR (red1)", *J. Am. Chem. Soc.,* vol. 123, no. 25, pp. 5853-5860, 2001.
[http://dx.doi.org/10.1021/ja010428d] [PMID: 11414817]

[59]   J. Yan, A.D. Kline, H. Mo, M.J. Shapiro, and E.R. Zartler, "The effect of relaxation on the epitope mapping by saturation transfer difference NMR", *J. Magn. Reson.,* vol. 163, no. 2, pp. 270-276, 2003.
[http://dx.doi.org/10.1016/S1090-7807(03)00106-X] [PMID: 12914842]

[60]   M. Mayer, and T.L. James, "NMR-based characterization of phenothiazines as a RNA binding scaffold", *J. Am. Chem. Soc.,* vol. 126, no. 13, pp. 4453-4460, 2004.
[http://dx.doi.org/10.1021/ja0398870] [PMID: 15053636]

[61]   Y.S. Wang, D. Liu, and D.F. Wyss, "Competition STD NMR for the detection of high-affinity ligands and NMR-based screening", *Magn. Reson. Chem.,* vol. 42, no. 6, pp. 485-489, 2004.
[http://dx.doi.org/10.1002/mrc.1381] [PMID: 15137040]

[62]   B. Meyer, and T. Peters, *NMR spectroscopy techniques for screening and identifying ligand binding to protein receptors.* vol. 42. International Edition. Angewandte Chemie, 2003, pp. 864-890.

[63]   H. Coelho, T. Matsushita, G. Artigas, H. Hinou, F.J. Cañada, R. Lo-Man, C. Leclerc, E.J. Cabrita, J. Jiménez-Barbero, S. Nishimura, F. Garcia-Martín, and F. Marcelo, "The quest for anticancer vaccines: Deciphering the fine-epitope specificity of cancer-related monoclonal antibodies by combining microarray screening and saturation transfer difference NMR", *J. Am. Chem. Soc.,* vol. 137, no. 39, pp. 12438-12441, 2015.
[http://dx.doi.org/10.1021/jacs.5b06787] [PMID: 26366611]

[64]   A. Viegas, J. Manso, M.C. Corvo, M.M. Marques, and E.J. Cabrita, "Binding of ibuprofen, ketorolac, and diclofenac to COX-1 and COX-2 studied by saturation transfer difference NMR", *J. Med. Chem.,* vol. 54, no. 24, pp. 8555-8562, 2011.
[http://dx.doi.org/10.1021/jm201090k] [PMID: 22091869]

[65]   M.S. Estevão, L.C.R. Carvalho, M. Freitas, A. Gomes, A. Viegas, J. Manso, S. Erhardt, E. Fernandes, E.J. Cabrita, and M.M. Marques, "Indole based cyclooxygenase inhibitors: synthesis, biological evaluation, docking and NMR screening", *Eur. J. Med. Chem.,* vol. 54, pp. 823-833, 2012.
[http://dx.doi.org/10.1016/j.ejmech.2012.06.040] [PMID: 22796043]

[66]   M. Mayer, and B. Meyer, "Group epitope mapping by saturation transfer difference NMR to identify segments of a ligand in direct contact with a protein receptor", *J. Am. Chem. Soc.,* vol. 123, no. 25, pp. 6108-6117, 2001.
[http://dx.doi.org/10.1021/ja0100120] [PMID: 11414845]

[67]   M. Vogtherr, and T. Peters, "Application of NMR based binding assays to identify key hydroxy groups for intermolecular recognition", *J. Am. Chem. Soc.,* vol. 122, no. 25, pp. 6093-6099, 2000.
[http://dx.doi.org/10.1021/ja0001916]

[68]   A. Ardá, P. Blasco, D. Varón Silva, V. Schubert, S. André, M. Bruix, F.J. Cañada, H.J. Gabius, C. Unverzagt, and J. Jiménez-Barbero, "Molecular recognition of complex-type biantennary N-glycans by protein receptors: a three-dimensional view on epitope selection by NMR", *J. Am. Chem. Soc.,* vol. 135, no. 7, pp. 2667-2675, 2013.
[http://dx.doi.org/10.1021/ja3104928] [PMID: 23360551]

[69]   T. Diercks, J.P. Ribeiro, F.J. Cañada, S. André, J. Jiménez-Barbero, and H.J. Gabius, "Fluorinated carbohydrates as lectin ligands: versatile sensors in 19F-detected saturation transfer difference NMR spectroscopy", *Chemistry,* vol. 15, no. 23, pp. 5666-5668, 2009.
[http://dx.doi.org/10.1002/chem.200900168] [PMID: 19388026]

[70]   R.S. Houliston, B.C. Jacobs, A.P. Tio-Gillen, J.J. Verschuuren, N.H. Khieu, M. Gilbert, and H.C. Jarrell, "STD-NMR used to elucidate the fine binding specificity of pathogenic anti-ganglioside antibodies directly in patient serum", *Biochemistry,* vol. 48, no. 2, pp. 220-222, 2009.
[http://dx.doi.org/10.1021/bi802100u] [PMID: 19105626]

[71]   A.J. Benie, R. Moser, E. Bäuml, D. Blaas, and T. Peters, "Virus-ligand interactions: identification and characterization of ligand binding by NMR spectroscopy", *J. Am. Chem. Soc.,* vol. 125, no. 1, pp. 14-15, 2003.
[http://dx.doi.org/10.1021/ja027691e] [PMID: 12515488]

[72]   S. Mari, D. Serrano-Gómez, F.J. Cañada, A.L. Corbí, and J. Jiménez-Barbero, "1D saturation transfer difference NMR experiments on living cells: the DC-SIGN/oligomannose interaction", *Angew. Chem. Int. Ed. Engl.,* vol. 44, no. 2, pp. 296-298, 2004.
[http://dx.doi.org/10.1002/anie.200461574] [PMID: 15614901]

[73]   B. Claasen, M. Axmann, R. Meinecke, and B. Meyer, "Direct observation of ligand binding to membrane proteins in living cells by a saturation transfer double difference (STDD) NMR spectroscopy method shows a significantly higher affinity of integrin alpha(IIb)beta3 in native platelets than in liposomes", *J. Am. Chem. Soc.,* vol. 127, no. 3, pp. 916-919, 2005.
[http://dx.doi.org/10.1021/ja044434w] [PMID: 15656629]

[74]   F. Marcelo, F. Garcia-Martin, T. Matsushita, J. Sardinha, H. Coelho, A. Oude-Vrielink, C. Koller, S. André, E.J. Cabrita, H.J. Gabius, S. Nishimura, J. Jiménez-Barbero, and F.J. Cañada, "Delineating binding modes of Gal/GalNAc and structural elements of the molecular recognition of tumor-associated mucin glycopeptides by the human macrophage galactose-type lectin", *Chemistry,* vol. 20, no. 49, pp. 16147-16155, 2014.
[http://dx.doi.org/10.1002/chem.201404566] [PMID: 25324212]

[75]   J. Angulo, I. Díaz, J.J. Reina, G. Tabarani, F. Fieschi, J. Rojo, and P.M. Nieto, "Saturation transfer difference (STD) NMR spectroscopy characterization of dual binding mode of a mannose disaccharide to DC-SIGN", *ChemBioChem,* vol. 9, no. 14, pp. 2225-2227, 2008.
[http://dx.doi.org/10.1002/cbic.200800361] [PMID: 18720494]

[76]   V. Jayalakshmi, and N. Rama Krishna, "CORCEMA refinement of the bound ligand conformation within the protein binding pocket in reversibly forming weak complexes using STD-NMR intensities", *J. Magn. Reson.,* vol. 168, no. 1, pp. 36-45, 2004.
[http://dx.doi.org/10.1016/j.jmr.2004.01.017] [PMID: 15082247]

[77]   N.R. Krishna, and V. Jayalakshmi, "Quantitative analysis of STD-NMR spectra of reversibly forming ligand-receptor complexes", *Top. Curr. Chem.,* vol. 273, pp. 15-54, 2008.

[http://dx.doi.org/10.1007/128_2007_144] [PMID: 23605458]

[78]   J. Angulo, P.M. Enríquez-Navas, and P.M. Nieto, "Ligand-receptor binding affinities from saturation transfer difference (STD) NMR spectroscopy: the binding isotherm of STD initial growth rates", *Chemistry,* vol. 16, no. 26, pp. 7803-7812, 2010.
[http://dx.doi.org/10.1002/chem.200903528] [PMID: 20496354]

[79]   W.S. Price, P.W. Kuchel, and B.A. Cornell, "Microviscosity of human erythrocytes studied with hypophosphite and 31P-NMR", *Biophys. Chem.,* vol. 33, no. 3, pp. 205-215, 1989.
[http://dx.doi.org/10.1016/0301-4622(89)80022-5] [PMID: 2804239]

[80]   W.S. Price, and L-P. Hwang, "Some recent decelopments in NMR approaches fo studying liquid molecular dynamics and their biological applications", *J. Chin. Chem. Soc. (Taipei),* vol. 39, no. 6, pp. 479-496, 1992.
[http://dx.doi.org/10.1002/jccs.199200082]

[81]   R. Huo, R. Wehrens, and J. Van Duynhoven, Assessment of techniques for DOSY NMR data processing.
[http://dx.doi.org/10.1016/S0003-2670(03)00752-9]

[82]   T. Brand, E.J. Cabrita, and S. Berger, "Intermolecular interaction as investigated by NOE and diffusion studies", *Prog. Nucl. Magn. Reson. Spectrosc.,* vol. 46, no. 4, pp. 159-196, 2005.
[http://dx.doi.org/10.1016/j.pnmrs.2005.04.003]

[83]   C.S. Johnson, "Effects of chemical exchange in diffusion-ordered 2D NMR spectra", *J Magn Reson Ser A [Internet],* vol. 102, pp. 214-218, 1993.http://www.sciencedirect.com/science/article/pii/S1064185883710934
[http://dx.doi.org/10.1006/jmra.1993.1093]

[84]   E.J. Cabrita, S. Berger, P. Bräuer, and J. Kärger, "High-resolution DOSY NMR with spins in different chemical surroundings: influence of particle exchange", *J. Magn. Reson.,* vol. 157, no. 1, pp. 124-131, 2002.
[http://dx.doi.org/10.1006/jmre.2002.2574] [PMID: 12202141]

[85]   A. Viegas, A.L. Macedo, and E.J. Cabrita, Ligand based nuclear magnetic resonance screening techniques.*Ligand Macromolecule Interactions in Drug Discovery, Methods in Molecular Biology.,* A.C.A. Roque, Ed., Springer: New York, USA, 2010, pp. 81-100.
[http://dx.doi.org/10.1007/978-1-60761-244-5_6]

[86]   M. Lin, M. J. Shapiro, and J. R. Wareing, "Diffusion-Edited NMR-Affinity NMR dor Direct Observation of Molecular Interactions," J. Am. Chem. Soc., vol. 119, pp. 5249–5250, 1997.

[87]   R.C. Anderson, M. Lin, and M.J. Shapiro, "Affinity NMR: decoding DNA binding", *J. Comb. Chem.,* vol. 1, no. 1, pp. 69-72, 1999.http://www.ncbi.nlm.nih.gov/pubmed/10746015 [Internet].
[http://dx.doi.org/10.1021/cc980004o] [PMID: 10746015]

[88]   C. Dalvit, and A. Vulpetti, "Technical and practical aspects of (19) F NMR-based screening: toward sensitive high-throughput screening with rapid deconvolution", *Magn. Reson. Chem.,* vol. 50, no. 9, pp. 592-597, 2012.
[http://dx.doi.org/10.1002/mrc.3842] [PMID: 22821476]

[89]   L.H. Lucas, and C.K. Larive, "Measuring ligand-protein binding using NMR diffusion experiments", *Concepts Magn. Reson. Part A Bridg. Educ. Res.,* vol. 20, no. 1, pp. 24-41, 2004.
[http://dx.doi.org/10.1002/cmr.a.10094]

[90]   Y. Cohen, L. Avram, and L. Frish, "Diffusion NMR spectroscopy in supramolecular and combinatorial chemistry: An old parameter—new insights", *Angew. Chemie - Int. Ed.,* vol. 44, pp. 520-554, 2005.

[91]   H. D. T. Mertens and D. I. Svergun, "Structural characterization of proteins and complexes using small-angle X-ray solution scattering.," J. Struct. Biol., vol. 172, no. 1, pp. 128–41, Oct. 2010.

[92]   R.P. Rambo, and J.A. Tainer, "Super-resolution in solution X-ray scattering and its applications to structural systems biology", *Annu. Rev. Biophys.,* vol. 42, pp. 415-441,

2013.http://www.ncbi.nlm.nih.gov/pubmed/23495971 [Internet].
[http://dx.doi.org/10.1146/annurev-biophys-083012-130301] [PMID: 23495971]

[93]   L. Boldon, F. Laliberte, and L. Liu, "Review of the fundamental theories behind small angle X-ray scattering, molecular dynamics simulations, and relevant integrated application.," Nano Rev., vol. 6, p. 25661, 2015.

[94]   J. Trewhella, "Small-angle scattering and 3D structure interpretation", *Curr. Opin. Struct. Biol.,* vol. 40, pp. 1-7, 2016.
[http://dx.doi.org/10.1016/j.sbi.2016.05.003] [PMID: 27254833]

[95]   A. Guinier, "La diffraction des rayons X aux tres petits angles : applications a l'etude de phenomenes ultramicroscopiques," Ann. Phys. (Paris)., no. 11. Sér. 12, pp. 161–237, 2010.

[96]   V.O. Glatter, and O. Kratky, "Small angle X-ray scattering", *Small Angle X-ray Scatt.,* vol. 36, no. 5, p. 1985, 1982.

[97]   O. Kratky, and G. Porod, "Roentgenuntersuchung Geloester Fadenmolekuele", *Recl. Trav. Chim. Pays Bas,* vol. 68, pp. 1106-1122, 1949.
[http://dx.doi.org/10.1002/recl.19490681203]

[98]   S. Doniach, "Changes in biomolecular conformation seen by small angle X-ray scattering", *Chem. Rev.,* vol. 101, no. 6, pp. 1763-1778, 2001.
[http://dx.doi.org/10.1021/cr990071k] [PMID: 11709998]

[99]   O. Glatter, "A new method for the evaluation of small-angle scattering data", *J. Appl. Cryst.,* vol. 10, no. 5, pp. 415-421, 1977.
[http://dx.doi.org/10.1107/S0021889877013879]

[100] P. Chacón, F. Morán, J.F. Díaz, E. Pantos, and J.M. Andreu, "Low-resolution structures of proteins in solution retrieved from X-ray scattering with a genetic algorithm", *Biophys. J.,* vol. 74, no. 6, pp. 2760-2775, 1998.
[http://dx.doi.org/10.1016/S0006-3495(98)77984-6] [PMID: 9635731]

[101] D.I. Svergun, V.V. Volkov, and M.B. Kozin, "New Developments in Direct Shape Determination from Small-Angle Scattering. 2. Uniqueness", *Acta Crystallogr. A,* vol. 52, no. 3, pp. 419-426, 1996.
[http://dx.doi.org/10.1107/S0108767396000177]

[102] D.I. Svergun, "Restoring low resolution structure of biological macromolecules from solution scattering using simulated annealing", *Biophys. J.,* vol. 76, no. 6, pp. 2879-2886, 1999.
[http://dx.doi.org/10.1016/S0006-3495(99)77443-6] [PMID: 10354416]

[103] D. Franke, and D.I. Svergun, "DAMMIF, a program for rapid ab-initio shape determination in small-angle scattering", *J. Appl. Cryst.,* vol. 42, no. Pt 2, pp. 342-346, 2009.
[http://dx.doi.org/10.1107/S0021889809000338] [PMID: 27630371]

[104] A. Grishaev, J. Wu, J. Trewhella, and A. Bax, "Refinement of multidomain protein structures by combination of solution small-angle X-ray scattering and NMR data", *J. Am. Chem. Soc.,* vol. 127, no. 47, pp. 16621-16628, 2005.
[http://dx.doi.org/10.1021/ja054342m] [PMID: 16305251]

[105] M. V Petoukhov, D. Franke, A. V Shkumatov, G. Tria, A. G. Kikhney, M. Gajda, C. Gorba, H. D. T. Mertens, P. V Konarev, and D. I. Svergun, "New developments in the ATSAS program package for small-angle scattering data analysis.," J. Appl. Crystallogr., vol. 45, no. Pt 2, pp. 342–350, Apr. 2012.

[106] E. Valentini, A.G. Kikhney, G. Previtali, C.M. Jeffries, and D.I. Svergun, "SASBDB, a repository for biological small-angle scattering data", *Nucleic Acids Res.,* vol. 43, no. Database issue, pp. D357-D363, 2015.
[http://dx.doi.org/10.1093/nar/gku1047] [PMID: 25352555]

[107] V. Lučič, A. Rigort, and W. Baumeister, "Cryo-electron tomography: the challenge of doing structural biology in situ", *J. Cell Biol.,* vol. 202, no. 3, pp. 407-419, 2013.
[http://dx.doi.org/10.1083/jcb.201304193] [PMID: 23918936]

[108] R.M. Glaeser, "How good can cryo-EM become?", *Nat. Methods,* vol. 13, no. 1, pp. 28-32, 2016.
[http://dx.doi.org/10.1038/nmeth.3695] [PMID: 26716559]

[109] J.L.S. Milne, M.J. Borgnia, A. Bartesaghi, E.E. Tran, L.A. Earl, D.M. Schauder, J. Lengyel, J. Pierson, A. Patwardhan, and S. Subramaniam, "Cryo-electron microscopy--a primer for the non-microscopist", *FEBS J.,* vol. 280, no. 1, pp. 28-45, 2013.
[http://dx.doi.org/10.1111/febs.12078] [PMID: 23181775]

[110] E.V. Orlova, and H.R. Saibil, "Structural analysis of macromolecular assemblies by electron microscopy", *Chem. Rev.,* vol. 111, no. 12, pp. 7710-7748, 2011.
[http://dx.doi.org/10.1021/cr100353t] [PMID: 21919528]

[111] E.E.H. Tran, M.J. Borgnia, O. Kuybeda, D.M. Schauder, A. Bartesaghi, G.A. Frank, G. Sapiro, J.L. Milne, and S. Subramaniam, "Structural mechanism of trimeric HIV-1 envelope glycoprotein activation", *PLoS Pathog.,* vol. 8, no. 7, p. e1002797, 2012.
[http://dx.doi.org/10.1371/journal.ppat.1002797] [PMID: 22807678]

[112] Y. Xiong, "From electron microscopy to X-ray crystallography: molecular-replacement case studies", *Acta Crystallogr. D Biol. Crystallogr.,* vol. 64, no. Pt 1, pp. 76-82, 2008.http://www.pubmedcentral.nih.gov/articlerender.fcgi?artid=2394795&tool=pmcentrez&rendertype=abstract [Internet].
[http://dx.doi.org/10.1107/S090744490705398X] [PMID: 18094470]

[113] D.H. Thomas, A. Rob, and D.W. Rice, "A novel dialysis procedure for the crystallization of proteins", *Protein Eng.,* vol. 2, no. 6, pp. 489-491, 1989.
[http://dx.doi.org/10.1093/protein/2.6.489] [PMID: 2710785]

[114] F.R. Salemme, "A free interface diffusion technique for the crystallization of proteins for x-ray crystallography", *Arch. Biochem. Biophys.,* vol. 151, no. 2, pp. 533-539, 1972.
[http://dx.doi.org/10.1016/0003-9861(72)90530-9] [PMID: 4625692]

[115] I.R. Krauss, A. Merlino, and A. Vergara, "An overview of biological macromolecule crystallization. Vol. 14", *Int. J. Mol. Sci.,* vol. 14, no. 6, pp. 11643-11691, 2013.
[http://dx.doi.org/10.3390/ijms140611643]

[116] H. Hope, "Cryocrystallography of biological macromolecules: a generally applicable method", *Acta Crystallogr. B,* vol. 44, no. Pt 1, pp. 22-26, 1988.
[http://dx.doi.org/10.1107/S0108768187008632] [PMID: 3271102]

[117] J.W. Pflugrath, "Practical macromolecular cryocrystallography", *Acta Crystallogr. F Struct. Biol. Commun.,* vol. 71, no. Pt 6, pp. 622-642, 2015.
[http://dx.doi.org/10.1107/S2053230X15008304] [PMID: 26057787]

[118] "ISRDB Cryoprotectant database for protein crystals." [Online]. Available: http://web.archive.org/web/20111011202903/http://idb.exst.jaxa.jp/db_data/protein/search-.

[119] A.L. Patterson, "A fourier series method for the determination of the components of interatomic distances in crystals", *Phys. Rev.,* vol. 46, no. 5, pp. 372-376, 1934.
[http://dx.doi.org/10.1103/PhysRev.46.372]

[120] G.L. Taylor, "Introduction to phasing", *Acta Crystallogr. D Biol. Crystallogr.,* vol. 66, no. Pt 4, pp. 325-338, 2010.
[http://dx.doi.org/10.1107/S0907444910006694] [PMID: 20382985]

[121] M.G. Rossmann, and D.M. Blow, "The detection of sub-units within the crystallographic asymmetric unit", *Acta Crystallogr.,* vol. 15, no. 1, pp. 24-31, 1962.
[http://dx.doi.org/10.1107/S0365110X62000067]

[122] A.J. McCoy, R.W. Grosse-Kunstleve, P.D. Adams, M.D. Winn, L.C. Storoni, and R.J. Read, "Phaser crystallographic software", *J. Appl. Cryst.,* vol. 40, no. Pt 4, pp. 658-674, 2007.
[http://dx.doi.org/10.1107/S0021889807021206] [PMID: 19461840]

[123] J.C. Cobas, and F.J. Sardina, "Nuclear magnetic resonance data processing. MestRe-C: A Software package for desktop computers. Vol. 19", *Concepts Magn. Reson. Part A Bridg. Educ. Res.,* pp. 80-96, 2003.
[http://dx.doi.org/10.1002/cmr.a.10089]

[124] R. Keller, *The Computer Aided Resonance Assignment Tutorial.* Goldau, Switzerland: Cantina Verlag. 2004. 1-81.

[125] W.F. Vranken, W. Boucher, T.J. Stevens, R.H. Fogh, A. Pajon, M. Llinas, E.L. Ulrich, J.L. Markley, J. Ionides, and E.D. Laue, "The CCPN data model for NMR spectroscopy: development of a software pipeline", *Proteins,* vol. 59, no. 4, pp. 687-696, 2005.
[http://dx.doi.org/10.1002/prot.20449] [PMID: 15815974]

[126] T. D. Goddard and D. G. Kneller, "SPARKY 3." University of California, San Francisco, 2008.

[127] M. Piotto, V. Saudek, and V. Sklenár, "Gradient-tailored excitation for single-quantum NMR spectroscopy of aqueous solutions", *J. Biomol. NMR,* vol. 2, no. 6, pp. 661-665, 1992.
[http://dx.doi.org/10.1007/BF02192855] [PMID: 1490109]

[128] T-L. Hwang, and A.J. Shaka, "Water supression that works. Excitation sculpting using arbitrary waveforms and pulsed filed gradients", *J. Magn. Reson. A,* vol. 112, pp. 275-279, 1995.
[http://dx.doi.org/10.1006/jmra.1995.1047]

[129] E. Prost, P. Sizun, M. Piotto, and J.M. Nuzillard, "A simple scheme for the design of solvent-suppression pulses", *J. Magn. Reson.,* vol. 159, no. 1, pp. 76-81, 2002.
[http://dx.doi.org/10.1016/S1090-7807(02)00003-4] [PMID: 12468306]

[130] A.G. Palmer, J. Cavanagh, and P.E. Wright, "Sensitivity improvement in proton-detected two-dimensional heteronuclear correlation NMR spectroscopy", *J. Magn. Reson.,* vol. 93, no. 1, pp. 151-170, 1991.

[131] L.E. Kay, P. Keifer, and T. Saarinen, "Pure absorption gradient enhanced heteronuclear single quantum correlation spectroscopy with improved sensitivity", *J. Am. Chem. Soc.,* no. 10, pp. 10663-10665, 1992.
[http://dx.doi.org/10.1021/ja00052a088]

[132] J. Schleucher, M. Schwendinger, M. Sattler, P. Schmidt, O. Schedletzky, S.J. Glaser, O.W. Sørensen, and C. Griesinger, "A general enhancement scheme in heteronuclear multidimensional NMR employing pulsed field gradients", *J. Biomol. NMR,* vol. 4, no. 2, pp. 301-306, 1994.
[http://dx.doi.org/10.1007/BF00175254] [PMID: 8019138]

[133] S. Grzesiek, and A. Bax, "The importance of not saturating H2O in protein NMR. Application to sensitivity enhancement and NOE measurements", *J. Am. Chem. Soc.,* no. 12, pp. 12593-12594, 1993.http://pubs.acs.org/doi/abs/10.1021/ja00079a052 [Internet].
[http://dx.doi.org/10.1021/ja00079a052]

[134] A. Viegas, J. Manso, F.L. Nobrega, and E.J. Cabrita, "Saturation-transfer difference (STD) NMR : A simple and fast method for ligand screening and characterization of protein binding", *J. Chem. Educ.,* vol. 88, pp. 990-994, 2011.
[http://dx.doi.org/10.1021/ed101169t]

[135] D. Wu, A. Chen, and C.S. Johnoson Jr, "An improved diffusion-ordered spectroscopy experiment incorporating bipolar-gradient pulses", *J. Magn. Reson. A,* vol. 115, pp. 260-264, 1995.
[http://dx.doi.org/10.1006/jmra.1995.1176]

[136] H.B. Stuhrmann, "Neutron small-angle scattering of biological macromolecules in solution", *J. Appl. Cryst.,* vol. 7, no. 1, pp. 173-178, 1974.
[http://dx.doi.org/10.1107/S0021889874009071]

[137] M.V. Petoukhov, and D.I. Svergun, "Joint use of small-angle X-ray and neutron scattering to study biological macromolecules in solution", *Eur. Biophys. J.,* vol. 35, no. 7, pp. 567-576, 2006.http://www.ncbi.nlm.nih.gov/pubmed/16636827 [Internet].

[http://dx.doi.org/10.1007/s00249-006-0063-9] [PMID: 16636827]

[138] D.A. Jacques, and J. Trewhella, "Small-angle scattering for structural biology--expanding the frontier while avoiding the pitfalls", *Protein Sci.,* vol. 19, no. 4, pp. 642-657, 2010.
[http://dx.doi.org/10.1002/pro.351] [PMID: 20120026]

92   *Essential Techniques for Medical and Life Scientists, Part 1*, 2018, 92-108

CHAPTER 3

Isothermal Titration Calorimetry

Ozlem Ustun Aytekin[1], Elvan Yilmaz Akyuz[1], Banu Bayram[1], Esen Tutar[2], Halime Hanım Pence[3] and Yusuf Tutar[1,4,*]

[1] *Health Sciences Faculty, Nutrition and Dietetics Department, University of Health Sciences 34668, Istanbul, Turkey*

[2] *Kahramanmaraş Sütçü Imam University, Science and Letters Faculty, Avsar Campus, 46060, Kahramanmaras, Turkey*

[3] *Faculty of Medicine, Division of Biochemistry, University of Health Sciences 34668, Istanbul, Turkey*

[4] *Faculty of Pharmacy, Division of Biochemistry, University of Health Sciences 34668, Istanbul, Turkey*

Abstract: Isothermal titration calorimetry (ITC) is a versatile and easy to use technique to study macromolecular interactions. Macromolecules interact with each other to perform biochemical functions. These interactions are commonly based on protein-ligand binding interactions and may occur among peptide, DNA, RNA, lipid, carbohydrate, ion, antibody, and enzymes. ITC experiments are relatively easy to perform, and no labelling is required for the set up. Performing an ITC experiment not only gives binding parameter but also it provides thermodynamic data. Further, the thermodynamic data provides nature of the interaction at molecular level *i.e.* hydrogen bonding, hydrophobic interactions.

Cooperativity of ligands and binding specificity of each site may be determined by this technique. Molecular weight range, number of interacting molecules, and optical clarity limit most of the instrumental techniques but ITC is not affected by these limitations. All these benefits make ITC a common method to determine binding constants. A combination of the technique with complementary methods augments its benefits in detecting molecular mechanism.

Keywords: Binding constant, Dissociation constant, Drug design, Entropy, Enthalpy, Food constituent analysis, Heat capacity, Protein-protein interaction, Protein-receptor interaction, Stoichiometry.

* **Corresponding author Yusuf Tutar:** University of Pharmacy, Department of Basic Pharmaceutical Sciences, Division of Biochemistry, Istanbul, Turkey; Tel: +90 216 418 96 16; E-mail: ytutar@outlook.com

INTRODUCTION

An essential concept about macromolecular interaction is to determine the strength of specific binding between interacting systems. Although several techniques exist, ITC offers several advantages over other methods.

These advantages determine binding constant, enthalpy and heat capacity change of binding ΔC_p, abbreviation, in a single run. This last parameter provides insights into the binding nature of the interactions. A negative value means hydrophobic interactions are formed, and a positive value means hydrophobic interactions are perturbed upon binding. ITC measures interaction's heat effect without labeling or modifying macromolecules. The method is also not restricted to macromolecular weight limit, number of macromolecules, and optical clarity but must be performed in-solution. Heat exchange of the interaction (endothermic or exothermic) helps in the calculation of binding constant and binding enthalpy. Therefore ITC can determine stability, strength, specificity, and stoichiometry of the biological interactions readily. The binding studies not only determine the number of binding sites but, may also detect the affinity of ligand for each binding site (Fig. **1**). Further, dependency among ligand binding sites and affinity modulation by other molecules may easily be monitored. Bound and unbound ligands at equilibrium and reaching to the equilibrium can be judged from the titration curve. This rigorous method may accurately characterize interactions between proteins, peptides, antibodies, lipids, carbohydrates, nucleic acid, small molecules, and drug molecules. ITC measures nanomolar to millimolar range affinities between molecules. However, at weak affinity, cautions must be taken for solubility and instrumental noise limitations for reliable data acquitions in ITC measurements.

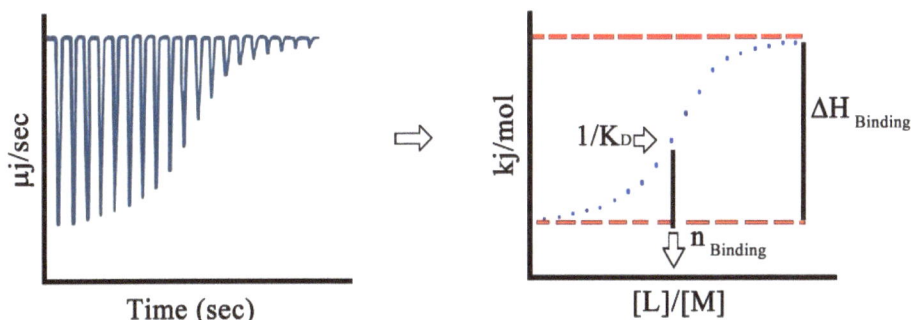

Fig. (1). Hypothetical ITC signal of ligand binding to a macromolecule (Left). The integral heat of interactions can be normalized per mole of ligand (Right). Thus, dissociation constant, binding stoichiometry, and enthalpy can be determined.

It is advisable to do two titrations: constant protein *versus* varying ligand concentration and constant ligand *versus* varying protein concentration. If the two titration curves overlap, this means that the stoichiometry is 1:1 between the macromolecules and the ligand. However, it is difficult to determine which concentration to start with if the nature of interaction is unpredictable [1].

Thermodynamic parameters can be determined since classical Gibbs free energy difference can be related to dissociation constant (Eq. 1 and 2).

$$\Delta G = -RT\ln K_D = \Delta H - T\Delta S \qquad (1)$$

$$K_D: 1/K_B = \frac{[L][M]}{[ML]} \qquad (2)$$

Where K_D is the dissociation and K_B is the binding constant. Using the thermodynamic parameters, nature of interaction may be elucidated. More negative ΔG means higher affinity, ΔH is a measure of broken and created bonds and this means that energy mainly originates from hydrogen and van der Waals bonds. A negative ΔH value favors binding however; a positive value of ΔS indicates entropically driven reactions. Further, $T\Delta S$ changes indicate alterations in hydrophobic interactions as well as in conformational changes.

The rest of the paper elaborates applications of ITC in the field of macromolecular interaction, drug research, and food technology.

3.1. ITC in Protein Science and Enzyme Technology

ITC is a convenient method to determine protein-protein, protein-DNA and protein-RNA interantions. ITC may also be employed in determining enzyme kinetics. Overall, ITC provides a versatile platform to characterize all protein interactions with other molecules.

3.1.1. Protein-protein Interactions

Protein-protein interactions are essential for cellular processes including gene regulation, cell signaling, and cell differentiation. Binding of two proteins causes changes in the thermodynamic potentials (ΔG, ΔH, ΔS). These parameters can readily be measured by calorimetry. ITC technique does not require any modification in proteins during measurements but other techniques (*i.e.*, surface plasmon resonance, analytical ultracentrifugation) require either further modification in the protein structure or the experiment takes longer time. Further, modification of the protein may alter the result of binding parameters and the other methods only provide binding parameter, however, ITC not only measures enthalpy, entropy values and binding affinity, but also provides a perspective for alterations of weak and hydrophobic interactions during binding processes. Therefore, folding pattern and structure of protein after interaction may be elucidated by the ITC measurements.

ITC consists of two calorimetric cells and ITC measurements compare heat differences of these cells after consequential injections of one reagent into one cell. This titration process compares protein-ligand interaction to protein alone. The reaction can either be exothermic and endothermic processes and ligand can be any macromolecule or can be more than one macromolecule. Analysis of this reaction, heat versus concentration of the titrating ligand, gives binding affinity and thermodynamic parameters at any point of the interaction.

Protein can interact with another homo- or heteroprotein. Even though, the heterodimeric or homodimeric complex experimentally differs in methodology, calculation of thermodynamic principles are similar in both situations [2].

Each injection corresponds to a new state and each state is associated with dependent variable, heat-qi. Using Eq.2 and Eq.3, thermodynamic parameters Ka, ΔHa, and ΔSa can be calculated.

$$q_i = V\Delta H_a([M_1 M_2]_i - [M_1\ M_2]_{i-1}) \tag{3}$$

M_1 and M_2 are the concentrations of each reactant [3]. In general, heterodimeric complex has been studied by ITC experiments to determine binding between two macromolecules, but the self -association of a homodimeric protein leading to the formation of homodimeric complex has usually not been studied by ITC.

Vieyra-Eusebio and Costas measured the interaction enthalpy between human serum albumin and lysozyme; non-functionally related proteins using ITC and found that the long-range electrostatic interactions, attractive or repulsive, due to the charges on the protein surfaces dominate the interaction between them, and that the strength of these solvent-mediated forces can be tuned by changing the pH [4]. In another study, the effect of different chloride concentration on ferredoxin and sulfite reductase interaction was studied by ITC [5]. Protein-protein interactions between the complex form indicated that intermolecular interactions mainly consist of electrostatic forces. To understand the interaction in molecular details, ITC data may be complemented by NMR to gain more insight into the interaction surfaces and binding affinity. Identification of interacting residues between protein-protein or protein-ligand as well as determining binding affinity may help researchers to elucidate optimum binding conditions. For example, changing buffer conditions to optimize effective electrostatic and ion interactions can enhance binding affinity [5].

3.1.2. Protein- DNA/RNA Interactions

ITC measurements are independent of macromolecule size and type therefore, not only protein-protein interaction can be measured but also, RNA-protein or DNA-protein interactions can be determined. ITC measurements are also independent of buffer type since the signal coming from buffer does not interfere with the signal. However, in spectroscopic techniques, signal from buffer may interfere with the signal. For example, circular dichroism signal of several proteins may alter with the content of the buffer employed. Further, ITC measurements do not require labeling or immobilization of interacting partners [6]. All these advantages help the direct measurement of ΔH, ΔS, and stoichiometry as well as K_d. RNA-protein interaction measurements of the complex formation, of *Caenorhabditis elegans* GLD-1 STAR domain and TGE RNA and the binding of *Aquifex aeolicus* S6:S18 ribosomal protein heterodimer to an S15-rRNA complex [6, 7] provided the nature of the binding mode of these two different macromolecules. The data analysis of the study showed that the number of binding sites indicates several experimental errors, if the numeric value deviates from 0.9–1.1 for 1:1 and 1.8–2.2 for 2:1 stoichiometry. If n<1, this means that the concentration of RNA has been overestimated or the concentration of protein has been underestimated, or another possibility is that not all of the RNAs are in an active state. In case of n>1, the concentration of protein and RNA have been underestimated or not all the proteins are active.

Binding of insulin to insulin-linked polymorphic region consensus DNA sequence was determined by ITC. The temperature dependent analysis ranging from 10 to 37 °C was made to determine K_D as well as ΔH, ΔG, and ΔS parameters. The binding was mainly driven by entropic factors below 25 °C however; driving parameter above 30 °C was enthalpic in nature. Further, ΔH was dependent on the ionization enthalpy of the phosphate buffer used indicating a proton release upon DNA binding to insulin. Insulin β subunit binds to polymorphic region of the DNA and large negative change in heat capacity for this interaction may be associated with hydrophobicity of this subunit amino acid sequence [8].

3.1.3. Enzyme Kinetics

ITC not only measures thermodynamic data but also kinetic rate of a reaction [9]. Kinetic rates at opaque solutions, suspensions, and multiple phase systems may easily be measured by heat alterations of the chemical reactions. Since heat is independent of light dependent signal, ITC has an advantage over conventional instrumental techniques.

The Michaelis–Menten/Briggs–Haldane (Eq.4) equation describes enzyme-derived kinetics [9].

$$dS/dt = -k_2[E_T][S]/(K_M + [S]) \tag{4}$$

[S] is the free substrate concentration, [ET] is the total enzyme concentration, k_2 is the rate constant, and K_M is the Michaelis–Menten, constant. The kinetics by ITC can be monitored by single and multiple injection methods depending on either enzyme or substrate as a titrant. Substrate injection into enzyme solution forms the basis of multiple injections since several titrant-substrates are injected and steady-state rate is measured after each titrant addition. Heat rate changes over substrate concentration after 20-40 points provide kinetic mechanism. The important part of this assay is to adjust enzyme and substrate concentrations to give measurable heat rate changes after each injection [10]. Thus, the rate of substrate reaction from Eq. 5 can be used to calculate $[S]_t$. K_M and k_2 or k_{cat} can be derived from Michaelis-Menten kinetics [10].

$$\left(\frac{dS}{dt}\right) = \frac{\frac{dQ_r}{dt}}{-V\Delta H_r} = k_2[E_T][S]/(K_M + [S]) \tag{5}$$

$$(dS/dt)_t = (dQ_{reaction}/dt)_t/(-\Delta_r HV) = k_2[E_T][S_t]/(K_M + [S]_t) \tag{6}$$

Heat rate of an exothermic reaction (positive Q and negative ΔH) is given in Eq. 5,6 where ΔH is the enthalpy change for the reaction (r), V is the volume of the ITC cell.

Single-injection method is just the reverse of multiple injection. A single enzyme injection into substrate solution and measurement of heat rate over time, typically 30 minutes provide kinetics of the reaction. However, data analysis is challenging in this assay due to uncertainties and approximations, but requires lower concentration of enzymes and faster than multiple injection method.

3.2. ITC in Pharmaceutical Research

ITC is one of the common methods in the pharmaceutical industry with its advantages over other well-known equipment. ITC is unique in providing quantitative thermodynamic information from a single titration and as mentioned earlier, the method is independent of molecular weight limitation and does not require any form of labelling of binding molecules [11, 12]. The instrument characteristics include rapid response and fast thermal equilibration, which can detect even some very week interactions [13].

ITC is widely used to study binding interactions. In the literature, many applications of ITC have been reported in pharmaceutical industry including drug delivery, drug discovery, drug screening, drug design and development, macromolecular assembly, interaction of surface to nanoparticles applications. These applications will be reviewed in this part of the chapter [14].

3.2.1. ITC in Drug Delivery

Drug degradation minimization, increased bioavailability and drug accumulation at the targeted area, optimize efficient drug delivery [15]. Drug partitioning in micellar systems lacks quantitative data in terms of energetics of the interactions. Micellar assemblies are widely used as drug transporting vehicles and interaction of drug with these vehicles as well as partitioning constant, stoichiometry and thermodynamic parameters (molar enthalpy and standard molar entropy change) are essential parameters for characterization of the complex. These parameters help determining the design of novel drugs, modification or choice of transport vehicles for drug delivery to specific targets [16].

Drug delivery into cell membranes is another essential biochemical process and ITC can be employed to measure interaction of drugs. Dual employment of ITC and fluorescence spectroscopy to determine thermodynamics of positively charged drug molecules (propranolol hydrochloride, tacrine and aminacrine), and negatively charged polymers mimicking cellular membrane to determine binding constant and thermodynamic parameters, highlight the importance of the formation of drug-membrane complex. The experiment is useful to define conditions for drug delivery into cell membranes [17].

Nanoparticle formulations require knowledge of drug loading, controlled drug release, stealthiness and drug targeting properties for an effective drug delivery system. Further, the administration of drugs and understanding biological effects require further knowledge of surface properties, affinities, and association between the properties of biological molecules with nanoparticles. ITC is useful to determine the interaction of the surface immobilized ligand of nanoparticles to the target site. High-binding affinity and strong selectivity of selected ligands are the basis of therapeutic and diagnostic innovations [11].

3.2.2. ITC in Drug Screening

As there are many candidate molecules for the treatment of metabolic disorders, it is required to select the appropriate one among the candidates. ITC may serve as a platform for the screening of drug candidates. Most drug discovery approaches use computational tools and modeling studies, which fail to detect slight details of molecular binding interactions. Wang *et al.* used ITC for the screening of drug candidates for Alzheimer's disease (AD) through bio-thermodynamic approach to real-time monitoring the heat of metabolism involved in the co-incubation of PC12 cells and Amyloid-beta (Aβ) [18]. One of the important progresses in AD is the deposition and aggregation of (Aβ) on cell membranes. In this study, the heat of metabolism associated with the effects of Aβ in different aggregation states on living cells was monitored by ITC. It was possible to measure the changes in` system's heat of cell culture through metabolic activity of living cells upon addition of Aβ species in various aggregation states. This approach allowed to know how fast a drug/chemical acts on cells and information about cell physiology [18].

3.2.3. ITC in Drug Design

In order to overcome poor water-solubility problems of the drugs, surfactants are commonly used in pharma industry. The complex drug-surfactant interactions can be better understood through quantitative thermodynamic data analysis. The enthalpy of the drug-surfactant interaction gives a valuable data for free energy transfer of the drug to each surfactant. Additionally, screening of surfactants for the solubility enhancement allows the selection of optimum solubilizing surfactants [11]. In some cases, high-tech equipment may become insufficient to reveal the enthalpic and entropic properties of packing in terms of weak electrostatic and water interactions and structural-dynamic changes. On the other hand, ITC readily provides thermodynamic data to profile optimal values for packing and designing innovative drugs [12].

ITC is one of the main techniques to determine the affinity between two molecules interacting in aqueous solution. This function is mostly used for protein-ligand interactions in drug design but it could be also used to measure interactions between any biochemical compound such as proteins, nucleic acids, lipids, carbohydrates, and other organic compounds. The most important advantage of ITC is that it is the only method that directly determines the enthalpy, which is a useful parameter for structure-thermodynamics correlations in the design of novel drug-like molecules [19].

The ADME (absorption, distribution, metabolism, and excretion) properties of drugs can be altered by serum albumin binding. Therefore, interaction mechanisms between drugs and this abundant serum protein are essential to understand the pharmacokinetics and pharmacodynamics of the drug since most of the drugs circulate in plasma to reach their target tissues. This interaction determines bioavailability and distribution of the drugs. Excretion rate of the drug affect the shelf life and therapeutic effect of the drug through modulating their delivery to the cells *in vivo* [20]. In this respect, the binding of potential drug candidates, functioning as biologically active molecules with many health promoting effects, namely catechins [19], ellegic acid [20] and proanthocyanidin B3 [21] to serum albumin was investigated. For beneficial therapeutic effects, metabolism and excretion of the molecules, protein binding affinity must be optimized [20, 22].

3.2.4. ITC in Drug Development

ITC experiments are of interest to drug development research. Understanding the thermodynamics of interactions of pharmaceuticals as functional ingredients, allows the choice of adequate processes for product quality optimization [14]. The first step upon drug development, is to perform shelf life tests and to determine the chemical and physical stability for the product quality. Characterization of active drug agents in raw and formulated forms determines the shelf-life of the product. Stability testing on pharmaceutical products can be tedious to perform because it requires analytical techniques that are specific to the product, for example, monitoring degradation rate of active ingredients requires an extraction/separation stage and then the analysis. However, in early stages of ingredients formulation, researchers need simple/robust methods for rapid formulation screening. ITC can fulfill such requirements with the provided information about product stability through sample thermal activity tracing [14]. Samples in ITC are not destructed and can be studied under storage conditions. These properties make the technique suitable for pharmaceutical applications. Thus, recovered samples can be used in other studies. This is important specially in the earlier phase of the drug design when synthesized samples are at milligram quantities [23].

The solid state of the pharmaceuticals (mechanical, physicochemical, textural properties) determines their physical stability. Amorphous materials produced by spray or freeze-drying processes are potentially valuable products. However, they may undergo significant physico-chemical changes during processing and storage. ITC is also frequently used for the stability prediction of amorphous products. Another critical point is to measure hydration and water vapor interactions for the shelf life determination [14].

3.3. ITC Applications in Food Science

Food constituents such as proteins, carbohydrates, lipids, surfactants, and mineral interactions can be measured by ITC conveniently. Macromolecule interactions are essential in the development of functional foods and improvement human health and well-being through nutritional treatments specially for cancer and diabetes patient treatment. During the process, ITC may provide qualitative and quantitative stability/compatibility studies. Several ITC ligand binding studies have been performed [14], especially between proteins and carbohydrates for characterizing biomolecular interactions in food industry. We may categorize ITC applications in food science under the topics below:

3.3.1. Interactions Involving Proteins or Peptides

Proteins and peptides may interact with several different types of ligands and ITC may detect binding constant and interaction stoichiometry in solution without further modifications of proteins or ligands in food science.

3.3.1.1. Interactions Between Flavan-3-ols and L-proline

Interactions of tannins with proteins is essential in astringency. Tannins interact with proteins as reported in the literature and this is essential in astringency. The characteristic feature is employed in protein fining procedures. Different flavan-3-ol monomers (catechin, epicatechin, epicatechin gallate, epigallocatechingallate) and an oligomeric form of grape seed tannins binding to poly(L proline) were determined by ITC. Although catechin and epicatechin did not bind to poly(L proline) as evidenced by no enthalpy change, large differences were determined in case of flavan-3-ol monomers. Further examination of ITC data indicated that the binding of grape seed tannins to proteins was cooperative [24].

3.3.1.2. Interactions Between Green Tea Flavanoids and Milk Proteins

Milk proteins interact with flavonoids and functional groups of other compounds in foods. As milk proteins have cross-linking property and affect the texture and nutritional properties, these interactions have gained interest among researchers. Polyphenols also react with proteins either by covalent or by weak binding. Beside their antioxidant capacity, green tea flavonoids (GT) interact with proteins and this is great interest to be revealed yet. GT catechins and protein interactions may alter their bioavailability and functionality.

Studies are performed for GT flavonoids binding to milk proteins. β-casein and (+) catechin interaction was determined by ITC to find out thermodynamic nature of the binding. The study revealed that milk protein system adapts to a conformation and exposes less hydrophobic regions on the outer surface upon GT flavonoid binding and protein-protein interaction [25]. Thus, ITC binding data not only display raw binding constant but also aids to show the nature of protein-ligand interaction.

3.3.1.3. Interactions Between Soy Protein and Acid Mineral Solution

Calcium carbonate is used for soy-based drink supplementation. Solubility of salt and precipitation of proteins are important characteristics which depend on physical-chemical stability of the system. Calcium carbonate has a neutral taste, on the contrary calcium chloride has a bitter taste. Therefore, calcium carbonate is preferred in supplemented products but it has lower solubility compared to that of calcium chloride. To increase the solubility of calcium carbonate in the presence of soy proteins citric acid was employed. Interactions of calcium carbonate, citric acid, and soy proteins were studied in solution by ITC to find the suitable formulation for the supplemented products [26].

3.3.1.4. Interactions Between Epicatechin and Serum Albumin

Binding of epicatechin to bovine serum albumin (BSA) was studied by ITC [27]. The study indicated that epicatechin interacts with BSA by non-covalent binding. Increase in BSA concentration led to change in interaction energetics, thus indicating epicatechin dependent BSA aggregation. Control experiments with hydrophobic proline rich gelatin did not interact with the protein suggesting that epicatechin based binding is mainly hydrogen bond driven rather than hydrophobic based interaction [28].

3.3.2. Interactions Involving Carbohydrates

Foods contain a variety of carbohydrates and these molecules have different molecular, physicochemical, and nutritional properties. Carbohydrates are also different in terms of chain length, branching, conformation, flexibility, polarity, and electrical properties. Further, their water-solubility, thickening properties, and gelation properties vary [29]. The studies on carbohydrates performed by ITC are as follow:

3.3.2.1. Carbohydrate Binding Properties of Banana Lectin

The interaction between lectins of banana (*Musa acuminata*) and the closely related plantain (Musa spp.) was monitored by ITC against monosaccharides, oligosaccharides, and their derivatives and banana lectins were found to determined to bind to mannose/glucose binding proteins. The banana lectin interacted with branched α-glucans and α-mannans but not with linear α-glucans containing only a-1,4 and a-1,6-linked glucose units [30].

3.3.2.2. Interactions Between Chitosan and Bile Salt

Bile acids binding to dietary fibers in the small intestine reduce blood cholesterol levels. Sodium taurocholate and cationic chitosan biopolymer binding was determined by ITC. To gain insight into the origin of the interaction, a range of salt concentration (0–150 mM NaCl) and different temperatures (10–40 °C) were tested. The temperature dependence of the reaction showed that hydrophobic interactions play an essential role during the binding process. ITC may provide origin and characteristic details of the interaction and as aforementioned, this information may help in the design of supplements and innovative drugs or functional foods [31].

3.3.3. Interactions Involving Lipids

Lipids in foods are present in dispersion of small oil droplets in aqueous medium. ITC has been employed to determine enthalpic changes of sodium dodecyl sulfate (SDS) micelles into silicone oil-in-water emulsions composed of droplets with different diameters. The study showed that SDS micelles augment oil droplets attraction and eventually this leading to flocculation after the concentration of SDS surfactant exceeds a certain concentration value. Thus, ITC technique provides insight into droplet–droplet interactions in emulsions and this measurement help in determining aggregation patterns of lipid droplets [29].

3.4. Protocol

Establishing a laboratory protocol requires a thorough knowledge and experience. Some key points are mentioned below to establish a procedure to successfully perform the ITC experiments.

Temperature set up is critical for the time management of the ITC experiments and hence, adjusting the desired temperature the day before performing the experiment ensures avoiding longer equilibration times. If a series of experiments planned, starting with the lowest temperature saves time since heating the instrument is faster than the cooling process. Further, periodic calibration of the instrument improves measurements and provides reproducible experiments [1].

Evaluation of ITC titration curve is based on binding dependent heat exchange process. Therefore, sample preparation and proper measurement of the binding reaction are two particularly essential steps for proper calorimetric titrations and for accurate and precise results.

3.4.1. Sample Preparation

Binding studies require pure protein (more than 95% purity) and the protein must be dialyzed against a suitable buffer. Further, the protein of interest must be dialyzed in the selected buffer and the last dialysis buffer must be used as a reference solution in the experiments. The ligand must be dissolved in the same buffer as well. Final salt concentration may perturb/prevent binding, thus if dialyses not possible, desalting columns may be used for desalting. After pH adjustment protein must be centrifuged and the buffer/ligand must be filtered to avoid any insoluble particles as well as degassing the solutions to avoid bubbles. The exact concentration of protein and ligand must be accurately determined.

Spectrophotometric methods by using extinction coefficient are common. However, ligand concentration may be determined by other techniques as well and this depends on the nature of the ligand. A convenient method may be selected *i.e.* metal ligand concentration can be determined by atomic absorption spectrophotometry. Furthermore, the calibration of pipettes by weight for concentration determination of the molecules is important for performing ITC experiments.

Buffers have ionization enthalpies (ΔH_{ion}) and buffers such as acetate, phosphate, formate, sulfate, citrate, and glycine almost have zero ionization enthalpies and are preferred for ITC experiments; however, buffers like quaternary amines such as Tris have high ionization enthalpies and are not preferred for titration calorimetry experiments.

Protein stability and solubility must also be considered when choosing a proper type of buffer. Also, 1,4 -dithiotreitol (DTT) should be avoided since it is unstable and undergoes oxidation. Instead of employing such aggregation inhibitors β-mercaptoethanol and Tris-(carboxyethyl)phosphine (TCEP) can be used in the buffer systems. It should be noted that TCEP is not stable in phosphate buffer.

3.4.2. Measurement

The cell must be washed several times with last dialysis buffer and the buffer should be removed from the cell but drying the cell should be avoided. Then, the cell can be filled with the protein solution carefully. After filling the syringe, air bubbles should be removed and not introduced into the cell. The cell must be filled until the reservoir is overflown. Thus, pumping solution in and out will remove trapped bubbles out of the cell. Special care must be taken for not bending the syringe since it will disturb the baseline of the instrument.

The reference cell generally contains 0.1% sodium azide dissolved in water and replaced each month.

Once the sample and ITC stirring syringe are loaded, titration program can be started. After the titration is complete, stirring syringe can be removed and sample cell must be washed with distilled water. Then, this ligand must be removed from stirring syringe and it must be washed with distilled water and air-dried with careful handling. Sample cell must be filled with water, and the solution must be replaced regularly.

3.4.3. Data Collection and Analysis

The data is collected by the help of instrument software. Once the instrument is equilibrated, the signal is stable. Then, the ligand injection to the sample cell can be started and the titration continues until the heat changes are not observed. This point indicates the ligand saturation of the protein. Protein-ligand and ligand only ligand titration experiments with ITC will allow binding analysis. The software allows data analysis to determine the number of binding sites and carryout cooperativity calculations.

3.5. Troubleshooting

K_D value is not known for several binding reactions and an experimental set up with expected value may not help in determining the K_D value experimentally. Therefore, a range of titrant concentration must be tested and a supportive set of experiments must be performed. For example, to determine binding at different protein concentrations for a range of ligand concentration and then, at fixed ligand concentrations, a set of protein titrations at a range of protein concentration must be performed to fully elucidate the number of binding site(s) and cooperativity accurately.

Other adverse situations for ITC experiments are set up that cause rapid saturation of binding sites, no binding, too little heat exchange, this never reaching to the saturation level. These can be corrected with proper experimental set up as stated in the previous paragraph. Further, increasing sample concentration and changing experimental temperature are other alternatives for solving the problems.

Baseline position is the first diagnosis for ITC experiments data quality. Air bubbles and bent syringe are common instrument based problems. Cell cleanliness, sticky proteins and time between injections may also lead to problems. Increasing the time between the injections may also provide better signal.

Furthermore, performing water-water titration may help understanding any instrumental problem and calibration of the instrument eliminates the errors.

CONSENT FOR PUBLICATION

Not applicable.

CONFLICT OF INTEREST

The author (editor) declares no conflict of interest, financial or otherwise.

ACKNOWLEDGEMENTS

Declare none.

REFERENCES

[1]   M.M. Lopez, and G.I. Makhatadze, "Isothermal titration calorimeter. Calcium-binding protein protocols: Methods and techniques of the series", *Methods Mol. Biol.,* vol. 173, no. 2, pp. 121-126, 2002.
         [PMID: 11859755]

[2]   A. Velazquez-Campoy, S.A. Leavitt, and E. Freire, "Characterization of protein-protein interactions by isothermal titration calorimetry", *Methods Mol Biol.,* vol. 261, pp. 35-54, 2004.
         [http://dx.doi.org/10.1385/1-59259-762-9:035] [PMID: 15064448]

[3]   N.M. Walavalkar, N. Gordon, and D.C. Williams Jr, "Unique features of the anti-parallel, heterodimeric coiled-coil interaction between methyl-cytosine binding domain 2 (MBD2) homologues and GATA zinc finger domain containing 2A (GATAD2A/p66α)", *J. Biol. Chem.,* vol. 288, no. 5, pp. 3419-3427, 2013.
         [http://dx.doi.org/10.1074/jbc.M112.431346] [PMID: 23239876]

[4]   M.T. Vieyra-Eusebio, and M. Costas, "Protein-protein interactions at high concentrations. Isothermal titration calorimetry determination of human serum albumin-lysozyme interaction enthalpy at several pH values", *Thermochim. Acta,* vol. 641, pp. 39-42, 2016.
         [http://dx.doi.org/10.1016/j.tca.2016.08.011]

[5]   J.Y. Kim, T. Ikegami, and Y. Goto, Investigation of protein-protein interactions of ferredoxin and sulfite reductase under different sodium chloride concentrations by NMR spectroscopy and isothermal titration calorimetry. *Molecular Physiology and Ecophysiology of Sulfur.* Springer International Publishing, 2015, pp. 169-177.
         [http://dx.doi.org/10.1007/978-3-319-20137-5_17]

[6]   M.I. Recht, S.P. Ryder, and J.R. Williamson, "Monitoring assembly of ribonucleoprotein Complexes by isothermal titration calorimetry." Methods Mol Biol, vol. 488, pp. 117-127, 2008.
         [http://dx.doi.org/10.1007/978-1-60327-475-3_8]

[7]   A.L. Feig, "Applications of isothermal titration calorimetry in RNA biochemistry and biophysics", *Biopolymers,* vol. 87, no. 5-6, pp. 293-301, 2007.
         [http://dx.doi.org/10.1002/bip.20816] [PMID: 17671974]

[8]   C.M. Timmer, N.L. Michmerhuizen, A.B. Witte, M. Van Winkle, D. Zhou, and K. Sinniah, "An isothermal titration and differential scanning calorimetry study of the G-quadruplex DNA-insulin interaction", *J. Phys. Chem. B,* vol. 118, no. 7, pp. 1784-1790, 2014.
         [http://dx.doi.org/10.1021/jp411293r] [PMID: 24459986]

[9]   L.D. Hansen, M.K. Transtrum, C. Quinn, and N. Demarse, "Enzyme-catalyzed and binding reaction kinetics determined by titration calorimetry", *Biochim. Biophys. Acta,* vol. 1860, no. 5, pp. 957-966,

2016.
[http://dx.doi.org/10.1016/j.bbagen.2015.12.018] [PMID: 26721335]

[10]   M.K. Transtrum, L.D. Hansen, and C. Quinn, "Enzyme kinetics determined by single-injection isothermal titration calorimetry", *Methods,* vol. 76, pp. 194-200, 2015.
[http://dx.doi.org/10.1016/j.ymeth.2014.12.003] [PMID: 25497059]

[11]   K. Bouchemal, "New challenges for pharmaceutical formulations and drug delivery systems characterization using isothermal titration calorimetry", *Drug Discov. Today,* vol. 13, no. 21-22, pp. 960-972, 2008.
[http://dx.doi.org/10.1016/j.drudis.2008.06.004] [PMID: 18617012]

[12]   K. Rajarathnam, and J. Rösgen, "Isothermal titration calorimetry of membrane proteins-Progress and challenges", *BBA,* vol. 1838, pp. 69-77, 2014.
[http://dx.doi.org/10.1016/j.bbamem.2013.05.023] [PMID: 23747362]

[13]   K. Bernaczek, A. Mielanczyk, and Z.J. Grzywna, "Interactions between fluorescein isothiocyanate and star-shaped polymer carriers studied by isothermal titration calorimetry (ITC)", *Thermochim. Acta,* vol. 641, pp. 8-13, 2016.
[http://dx.doi.org/10.1016/j.tca.2016.08.007]

[14]   N. Khalef, O. Campanella, and A. Bakri, "Isothermal calorimetry: methods and applications in food and pharmaceutical fields", *Curr. Opin. Food Sci.,* vol. 9, pp. 70-76, 2016.
[http://dx.doi.org/10.1016/j.cofs.2016.09.004]

[15]   S. Choudhary, P. Talele, and N. Kishore, "Thermodynamic insights into drug-surfactant interactions: Study of the interactions of naporxen, diclofenac sodium, neomycin, and lincomycin with hexadecytrimethylammonium bromide by using isothermal titration calorimetry", *Colloids Surf. B Biointerfaces,* vol. 132, pp. 313-321, 2015.
[http://dx.doi.org/10.1016/j.colsurfb.2015.05.031] [PMID: 26057731]

[16]   N. Zhang, P.R. Wardwell, and R.A. Bader, ""Polysaccharide based micelles for drug delivery"", *Pharmaceutics,* vol. 5, pp. 329-352, 2013.

[17]   H.A. Santos, J.A. Manzanares, L. Murtomäki, and K. Kontturi, "Thermodynamic analysis of binding between drugs and glycosaminoglycans by isothermal titration calorimetry and fluorescence spectroscopy", *Eur. J. Pharm. Sci.,* vol. 32, no. 2, pp. 105-114, 2007.
[http://dx.doi.org/10.1016/j.ejps.2007.06.003] [PMID: 17643273]

[18]   S.S.S. Wang, M.S. Lin, S.L. Chen, Y. Chang, R.C. Ruaan, and W.Y. Chen, "Using isothermal titration calorimetry to real-time monitor the heat of metabolism: a case study using PC12 cells and Aβ(1-40)", *Colloids Surf. B Biointerfaces,* vol. 83, no. 2, pp. 307-312, 2011.
[http://dx.doi.org/10.1016/j.colsurfb.2010.11.038] [PMID: 21190816]

[19]   V. Linkuvienė, G. Krainer, W.Y. Chen, and D. Matulis, "Isothermal titration calorimetry for drug design: Precision of the enthalpy and binding constant measurements and comparison of the instruments", *Anal. Biochem.,* vol. 515, pp. 61-64, 2016.
[http://dx.doi.org/10.1016/j.ab.2016.10.005] [PMID: 27717855]

[20]   X. Li, and Y. Hao, "Probing the binding of (+)-catechin to bovine serum albumin by isothermal titration calorimetry and spectroscopic techniques", *J. Mol. Struct.,* vol. 1091, pp. 109-117, 2015.
[http://dx.doi.org/10.1016/j.molstruc.2015.02.082]

[21]   R Pattanayak, P Basak, and S. Sen, "Interaction of KRAS G-quadruplex with natural polyphenols: A spectroscopic analysis with molecular modeling", *Int. J. Biol. Macromol,* vol 89, pp. 228-237, 2017.

[22]   X. Li, and Y. Yan, "Probing the binding of procyanidin B3 to human serum albumin by isothermal titration calorimetry", *J. Mol. Struct.,* vol. 1082, pp. 170-173, 2015.
[http://dx.doi.org/10.1016/j.molstruc.2014.11.020]

[23]   M.A.A. O'Neill, and S. Gaisford, "Application and use of isothermal calorimetry in pharmaceutical development", *Int. J. Pharm.,* vol. 417, no. 1-2, pp. 83-93, 2011.

[http://dx.doi.org/10.1016/j.ijpharm.2011.01.038] [PMID: 21277961]

[24]   C. Poncet-Legrand, C. Gautier, V. Cheynier, and A. Imberty, "Interactions between flavan-3-ols and poly(L-proline) studied by isothermal titration calorimetry: effect of the tannin structure", *J. Agric. Food Chem.,* vol. 55, no. 22, pp. 9235-9240, 2007.
[http://dx.doi.org/10.1021/jf071297o] [PMID: 17850090]

[25]   Z. Yuksel, E. Avci, and Y.K. Erdem, "Characterization of binding interactions between green tea flavonoids and milk proteins", *Food Chem.,* vol. 121, pp. 450-456, 2010.
[http://dx.doi.org/10.1016/j.foodchem.2009.12.064]

[26]   L.S. Canabady-Rochelle, C. Sanchez, M. Mellema, and S. Banon, "Calcium carbonate-hydrolyzed soy protein complexation in the presence of citric acid", *J. Colloid Interface Sci.,* vol. 345, no. 1, pp. 88-95, 2010.
[http://dx.doi.org/10.1016/j.jcis.2010.01.037] [PMID: 20129621]

[27]   A. Cooper, "Thermodynamic analysis of biomolecular interactions", *Curr. Opin. Chem. Biol.,* vol. 3, no. 5, pp. 557-563, 1999.
[http://dx.doi.org/10.1016/S1367-5931(99)00008-3] [PMID: 10508661]

[28]   R.A. Frazier, A. Papadopoulou, and R.J. Green, "Isothermal titration calorimetry study of epicatechin binding to serum albumin", *J. Pharm. Biomed. Anal.,* vol. 41, no. 5, pp. 1602-1605, 2006.
[http://dx.doi.org/10.1016/j.jpba.2006.02.004] [PMID: 16522360]

[29]   I.J. Arroyo-Maya, and D.J. McClements, "Application of ITC in foods: A Powerful tool for understanding the gastrointestinal fate of lipophilic compounds", *Biochim. Biophys. Acta,* vol. 1860, no. 5, pp. 1026-1035, 2016.
[http://dx.doi.org/10.1016/j.bbagen.2015.10.001] [PMID: 26456046]

[30]   H. Mo, H.C. Winter, E.J. Van Damme, W.J. Peumans, A. Misaki, and I.J. Goldstein, "Carbohydrate binding properties of banana (Musa acuminata) lectin I. Novel recognition of internal alpha1,3-linked glucosyl residues", *Eur. J. Biochem.,* vol. 268, no. 9, pp. 2609-2615, 2001.
[http://dx.doi.org/10.1046/j.1432-1327.2001.02148.x] [PMID: 11322880]

[31]   M. Thongngam, and D.J. McClements, "Isothermal titration calorimetry study of the interactions between chitosan and a bile salt (sodium taurocholate)", *Food Hydrocoll.,* vol. 19, pp. 813-819, 2005.
[http://dx.doi.org/10.1016/j.foodhyd.2004.11.001]

CHAPTER 4

Reverse Transcription Polymerase Chain Reaction (RT-PCR)

Venkidasamy Baskar and **Sathishkumar Ramalingam**[*]

Plant Genetic Engineering Laboratory, Department of Biotechnology, Bharathiar University, Coimbatore, Tamil Nadu, India

Abstract: RT-PCR is a rapid and highly sensitive method to study the gene expression. RT-PCR includes the reverse transcription of RNA into cDNA and PCR. In RT-PCR, RNA is used as a template rather than DNA. Total RNA or poly (A) RNA can be used as a template for cDNA synthesis. RT reactions include primers (random primers/oligo dT/gene specific primers), reverse transcriptase and reaction buffers. RT-PCR can be performed either in one step or two step formats. In one step RT-PCR, reverse transcription and PCR were carried out sequentially in a single tube. However, in two step RT-PCR, RT reaction was performed separately, and one tenth of the RT reaction is used for PCR. In order to perform the semi-quantitative gene expression analysis, RT-PCR results will be subjected to the densitometry scanning and image J software analysis. RT-PCR is a powerful technique used in the various fields such as medical diagnostics, forensics and gene cloning. In this chapter, we have described the methodology, applications and the possible troubleshooting guide to be carried out during the RT- PCR experiment.

Keywords: Complementary DNA, DNase, Microarray, Nanodrop spectrophotometer, Nuclease free, Oligo dT, qPCR, RNA, Random primers, Reverse transcriptase, RNase, Reverse transcription, RT-PCR, RNA-Seq, TRIzol.

INTRODUCTION

Polymerase chain reaction (PCR) was developed by Kary Mullis at the Cetus Corporation in 1983. PCR includes various steps such as, template denaturation, primer annealing and elongation of primer to amplify the DNA sequences [1]. Unlike PCR, reverse transcription PCR (RT-PCR), uses RNA as a starting material. Reverse transcription is a process in which RNA is transcribed into complementary DNA (cDNA) that serves as a template to amplify genes.

[*] **Corresponding author Sathishkumar Ramalingam:** Plant Genetic Engineering Laboratory, Department of Biotechnology, Bharathiar University, Coimbatore, Tamil Nadu, India; Tel: +91 422 2428299; E-mail: rsathish@buc.edu.in

Yusuf Tutar (Ed.)

Total RNA (mRNA, tRNA and rRNA) or poly A RNA is used as a template for RT-PCR. Reverse transcription reaction mixture includes primers (random primers, oligo dT or a gene specific primer), reverse transcriptase (RTase, RNA-dependent DNA polymerase enzyme) and RNase inhibitor.

The native RTase (a multifunctional enzyme) possess, 1) a weak RNA-dependent DNA polymerase activity that lacks the 3'-5' exonuclease activity and (2) an RNase H activity that degrades the RNA template from RNA-DNA hybrid strands as the cDNA synthesis proceeds. Two types of RTases such as, M-MLV RTase derived from the Moloney murine leukemia virus, and the AMV RTase from the avian myeloblastosis virus are the most commonly used RTases in molecular biological applications. AMV RTase and M-MLV RTase are more suitable for the synthesis of short and long size cDNAs respectively [2]. There are two types of RT-PCRs, such as one-step and two-step RT-PCR based on the reactions performed.

In one-step RT-PCR, both the reverse transcription and PCR are performed sequentially in a single tube. However, in two-step RT-PCR, reverse transcription (RT) reaction is performed at optimal conditions and 10% of the RT reaction is used for subsequent PCR reaction [3]. RT-PCR is a rapid and sensitive method to detect and quantify the gene expression, it works even with low quantities of RNA and exhibits higher specificity (selective amplification of the desired product) [4, 5]. It is highly sensitive and simple to perform as compared to other techniques such as, *in situ* hybridization, northern blots and S1 nuclease assays [6, 7]. In this chapter, the methodology, applications of RT-PCR and guidelines for troubleshooting there been explained in detail.

4.1. Methodology

RT-PCR starts with the isolation of RNA, conversion of RNA into cDNA (reverse transcription) and amplification of the desired product using PCR. Each step should be carried out carefully to obtain the desired results.

4.1.1. Isolation of RNA

RT-PCR requires a full-length RNA with high integrity. It should be free from DNA, inhibitors of reverse transcriptase such as EDTA or SDS [8]. Partially degraded RNA can be used in RT-PCR for qualitative gene expression studies. However, full-length, high-quality RNA is essential to determine the quantitative gene expression. The use of oligo dT is not as efficient as random primers to synthesize cDNA from partially degraded RNA. Oligo dT will be suitable for PCR targets near the 3' end of mRNA. Several methods were employed to extract RNA from diverse biological sources such as plants, microbes, and animals.

Guanidine isothiocyanate/acidic phenol protocol (single step method) and TRIzol reagent method have been widely employed for RNA isolation from various cells and tissues [9, 10]. TRIzol method is an effective method to isolate high-quality RNA, even with the smallest amount of samples (100 cells or 1 mg of tissue) [11]. In addition, commercial RNA isolation kits impart good yield and fine quality RNA. Nuclease-free or diethyl pyrocarbonate (DEPC) treated tubes, reagents and water should be used for RNA isolation in order to avoid RNA degradation. Here, we described the RNA isolation from the plant leaf samples using the TRIzol method.

1. Grind the plant leaf tissue (50-100 mg) into a fine powder mix it with TRIzol (1 ml) and leave at room temperature (RT) for 10 min.
2. Add 200 µl chloroform and shake vigorously for 10-15 sec. Allow the mixture to stand for 10 min at RT and centrifuge at 11000 rpm for 15 min at 4°C. Centrifugation separates the mixture into three phases such as 1) red organic phase containing protein, 2) an interphase possesses DNA, 3) upper aqueous phase containing RNA.
3. Transfer the aqueous phase without disturbing interphase into the fresh RNase-free tube and add 500 µl of isopropanol. Mix the sample followed by incubation at RT for 5-10 min.
4. Centrifuge the mixture at 11000 rpm for 10 min at 4°C.
5. RNA will form a pellet, which will be attached to the sides and bottom of the tube.
6. Wash the pellet with 75% ethanol (1 ml). Vortex and centrifuge at 6500 rpm for 5 min at 4°C.
7. Briefly, dry the RNA pellet for 5 min by air drying. Do not dry the RNA pellet completely, as it will reduce its solubility.
8. Add 50 µl DEPC water to dissolve the RNA. The dissolved RNA will be ready for cDNA synthesis.

Longer transcripts (>4kb) are highly susceptible to RNases as compared to shorter transcripts. RNase inhibitor treatment protects the RNA from RNase degradation. In addition, DNase I treatment is essential to eliminate the residual DNA. To maintain RNA integrity for the long term, RNA can be dissolved in deionized formamide and stored at -70°C [12]. Moreover, RNA extracted from the RNase-rich samples (*e.g.,* pancreas) should be stored in formamide to get rid of RNA degradation. The size and integrity of the purified RNA can be determined in formaldehyde denaturing gel electrophoresis by visual inspection, under UV trans illuminator after ethidium bromide staining. The purified RNA can be quantified with the UV spectrophotometer (A260). However, it needs large sample volume for analysis. On the other hand, nanodrop UV-Vis Spectrophotometer (Nanodrop ND-1000 Spectrophotometer; Celbio, Italy) requires just 1 µl sample to determine the RNA purity (A260/280 ratio) and concentration (A260).

4.1.2. Reverse Transcription or Complementary DNA Synthesis (cDNA Synthesis)

cDNA synthesis can be performed in one step or two steps. A two-step cDNA synthesis method is described below.

All the reagents and RNA are stored at 4°C.

Step-1

1. Prepare the following reaction mixture in a tube incubated at 4°C.

2. The concentration of RNA used for cDNA synthesis varies between 1 ng – 2 µg/reaction.

Reagents/Components		Volume	Final Concentration
RNA	Total RNA (or)	variable	1 ng - 2 µg/rxn
	poly (A) / mRNA	Variable	1 pg - 2 ng/rxn
Primers	Oligo dT (typically 12–18mers) (or)	1 µl	0.1 - 1 µM
	Random primers (nonomers) (or)	1 µl	10 µM
	Gene specific primers	variable	0.1 - 1 µM
dNTP mix (10 mM each)		1 µl	500 µM
Nuclease free water		up to 14.5 µl	-

Note: Oligo dT enables reverse transcription by targeting the 3' end of mRNA/ poly (A) RNA. Random primers utilize whole RNA populations (mRNA, rRNA, tRNA, and small nuclear RNAs) for reverse transcription.

3. Incubate the mixture at 65°C for 5 mins and keep it on ice for at least 1 min. Collect all the components by a brief centrifugation and add the following reagents:

Step-2 of cDNA Synthesis

Reagents	Volume	Final Concentration
5X RT Buffer	4 µl	1X
Ribonuclease Inhibitor (40 U/µl)	0.5 µl	20 U/rxn
Reverse transcriptase	1 µl	200 U/rxn

4. Mix all the components well and collect (20 µl) by a brief centrifugation. In the case of random primers used, the mixture should be incubated at 25°C for 10 mins

to facilitate its binding to the template RNA. This incubation step is not needed if oligo dT or gene specific primers are used.

5. Perform cDNA synthesis by incubating the tube at 42°C for 50 mins. The temperature and duration of the incubation vary depending on the type of reverse transcriptase used.

6. Stop the reaction by heating it at 85°C for 5 mins and then keep it on ice to cool down the mixture (RTase will inhibit the PCR, hence it must be heat inactivated). The newly synthesized first-strand cDNA is ready for immediate downstream applications, or it can be kept at -20°C for long-term storage.

4.1.3. PCR Amplification

4.1.3.1. Designing Primer

The success of the RT-PCR strictly depends on the primer designing and more attention should be given while designing for RT-PCR. Primer should be designed to the specific region of the cDNA to avoid the amplification of nonspecific region. Primers targeted to the intron region will result in the amplification of undesired size and it can also be used to detect the purity of the sample. In order to avoid the amplification of DNA fragments, primers can be designed to anneal the exon-exon boundary of the mRNA.

1. The ideal primer size should be 18-25 nucleotides in length.
2. The G/C content of the primer is also crucial since higher GC content leads to least denaturation during cycling and also prone to a non-specific interaction. For an ideal primer, G/C content should be 50%. The number of A/T and G/C should be in same ratio. The 3' terminal region of primers should be G or C but not T.
3. The melting temperature (Tm) should be in the range of 55-65°C.
4. The temperature differences between the pairs not to exceed 1-2°C.
5. The optimal size of the amplicon should be in the range of 100 – 150bp.
6. There are many online algorithms available to design primers for all kinds of PCR. PRIMER3 is the most commonly employed tool for designing primers.

4.1.3.2. PCR Amplification and Detection

One microlitre of cDNA is sufficient to perform RT-PCR. The PCR reaction mixture includes template cDNA, dNTPs, primers, appropriate buffers, Taq DNA polymerase. A typical amplification cycle consists of a denaturation step (95°C), a template–primer annealing step (42–60°C), and an extension step (72°C). The results of RT-PCR can be determined by visual inspection or densitometric quantification of bands after ethidium bromide staining. The visual evaluation of band intensity (qualitative method) is largely subject to variation caused by individual perception [13]. However, the application of densitometry scanning and image processing programs in RT-PCR aids in semi-quantitative determination. The optical density of each amplified band will be calculated using the image processing program (Image J) and numerically expressed as the relative density in comparison to the optical density of the background. In order to avoid the systemic errors, following factors such as PCR conditions, a number of amplification cycles, the thickness of the agarose gel, image capture and scanning procedures should be optimized. Moreover, to quantify the gene expression, all the results must be normalized to that of the housekeeping genes (*e.g.,* actin, tubulin, glyceraldehyde 3-phosphate dehydrogenase (*GAPD*), ubiquitin) that serve as an internal standard.

4.2. Applications

1. RT-PCR is a highly sensitive and an effective method, which can be used to detect and quantify even the rare transcripts that are present in limited amounts in the samples [14, 15].
2. In clinical diagnosis, RT-PCR is the most commonly used technique to detect pathogens, cancer and genetic diseases.
3. It is simple and easy to perform to study the expression of genes as compared to other quantitative techniques such as northern blot, S1 nuclease assays and *in situ* hybridization.
4. The RT-PCR method is used for semi-quantitative gene expression analysis when target mRNA levels are standardized with the reference mRNA levels (housekeeping genes).
5. In food processing industries, RT-PCR helps to detect the viable pathogens (bacteria, molds and yeasts) in the food sample [16].

4.3. Precautions to be Taken

1. All reagents, tubes, and distilled water to use for RNA isolation and RT-PCR should be nuclease-free.
2. To avoid genomic DNA contamination, RNA samples must be treated with the DNase I.
3. The integrity of RNA should be checked with the denaturing agarose gel electrophoresis.
4. The purity and the concentration of RNA should be determined by nanodrop UV-Vis Spectrophotometer.
5. RNA stocks and dilution must be prepared in DEPC treated water to avoid RNase-mediated degradation.
6. RNase H treatment is essential to remove the RNA from the RNA-DNA

hybrid, which increases the yield and length of the cDNA.

7. In the case of random oligomers used for cDNA synthesis, the reaction should be kept at 25°C for 10 minutes and then rose to 42°C for amplification.

8. The RT reaction temperature can be increased from 42°C to 50°C for GC-rich templates.

9. To minimize variations, all reactions should be performed at the same time and the reagents should be prepared as a master mix.

10. All reagents for cDNA synthesis must be thawed on ice and stored at -20°C immediately, after their use. Repeated thawing of reagents, chemicals cause decreased efficiency.

11. Pilot experiments should be carried out to find the optimal conditions (*e.g.,* annealing temperature, primer concentrations, cDNA concentration) for RT-PCR.

12. For semi-quantitative experiments, the expression of samples should be normalized to the expression of housekeeping genes (*e.g.,* actin, ubiquitin, tubulin, *GAPD*).

4.3.1. Real-time Quantitative PCR (RT-qPCR)

Real-time quantitative PCR (RT-qPCR) is the current generation technology where the changes in gene expression can be monitored in a real-time as the reaction progress through quantification of fluorescence signals produced by reporter dyes. It is more sensitive and reliable than the conventional RT-PCR or other gene expression analysis. Moreover, the gene expression can be measured absolutely with the help of RT-qPCR. SYBR green is the most widely used fluorescent dye in the RT-qPCR assays. The fluorescence dyes specifically bound to the double-stranded DNA and their intensity will be increased as the PCR cycle proceeds. The fluorescence signals detected in each PCR cycle was proportional to the initial copy number of the subjected and or target gene.

RT-qPCR is commonly employed for the diagnosis of diseases in clinical laboratories. Due to the dynamic range of efficiency RT-qPCR is the effective and desirable choice for the quantification of rare and ample genes in the given sample. The normal PCR and RT-PCR were most helpful for the detection and identification of the pathogens in the samples. However, RT-qPCR has the ability to detect viable or active pathogens in the samples due to the absolute and real-time quantification of gene expression changes. It is also used to quantify the pathogenic intensity in the clinical samples based on the gene expression analysis. Therefore, RT-qPCR is one of the important techniques in the routine clinical diagnosis. However, the success of the RT-qPCR is strictly depend on the control or reference genes selected for the gene normalization function. Moreover, the specific binding ability of fluorescent dyes to the double-stranded DNA leads to binding to all types of double-stranded DNA including the primer dimmers, which affects the sensitivity of the RT-qPCR.

4.4. Future Perspectives

Various molecular techniques have been widely employed for the gene expression analysis. The classical techniques such as northern blot, *in situ* hybridization, RNase protection assays and RT-PCR were used for the qualitative and semi-quantitative gene expression studies. Latest molecular techniques such as, real-time quantitative PCR (RT-qPCR), microarray, RNA sequencing (RNA-seq) analysis have been developed for the absolute quantification of gene expression.

Notes:

Problems	Solutions

Low or no yield of first strand cDNA	1. Ensure the integrity of the isolated RNA, by running in denaturing agarose gel electrophoresis.
	2. Replace the RNA-
	   Isolate intact mRNA or total RNA using commercial RNA purification kits.
	   RNA should be isolated in the presence of ribonuclease inhibitor. The reagents and labware must be free from RNase contamination.
	   An additional 70% ethanol wash helps to eliminate or reduce the RT inhibitors (SDS, EDTA, guanidinium chloride, formamide, Na_2PO_4, or spermidine).
	3. Increase the length of the RT incubation temperature (42°C) from 60 minutes to 90 minutes to facilitate the synthesis of cDNA from rare or long RNA targets.
	4. Increase the amount of template RNA.
	5. Ensure the cDNA synthesis primers (oligo dT, random primers or gene specific primers) are complementary to target sequence.
	6. If using random primers, incubate the reaction at 25°C for 10 minutes prior to increasing the temperature to 42°C for cDNA synthesis. This allows better annealing of random primers to RNA.
	7. Use oligo dT primer for eukaryotic RNA.

The molecular weight of the amplification product is higher than expected	1. RNA must be treated with RNase-free DNase I, to clarify the DNA contamination. 2. PCR primers should be designed to anneal the sequences of the exon-exon boundary of the target genes.
Multiple non-specific amplification products	1. The annealing temperature should be raised to reduce nonspecific amplification. 2. Ensure the primers are not self-complementary or complementary to each other. 3. Alternatively, use a longer primer.

Among them, RT-qPCR is a simple technique and extensively used for gene expression analysis due to its high accuracy and sensitivity. Relative (relative to other genes) and absolute (against a standard) quantification of gene expression can be measured using a real time PCR [17]. The expression patterns of a large number of genes can be measured simultaneously using microarray, which is based on the hybridization between oligo nucleotide probes (designed from known genes) on microarray platform and target genes in biological samples. It requires the prior information about the genes to be studied. A high-throughput latest technology called RNA sequencing used for comprehensive transcriptome study, which does not need the details of reference genome. Therefore, this technique is potentially useful to discover novel genes and non-coding RNAs [18]. The advent of these latest technologies one can easily determine the expression of thousands of genes simultaneously, with high accuracy even from the uncharacterized genome.

CONSENT FOR PUBLICATION

Not applicable.

CONFLICT OF INTEREST

The author (editor) declares no conflict of interest, financial or otherwise.

ACKNOWLEDGEMENTS

VB was supported by a grant (Sanction No. PDF/2016/000750) from the Department of Science and Technology, Science and Engineering Research Board, Government of India. Our group is also supported by UGC-SAP and DST-FIST funds.

REFERENCES

[1]   R.K. Saiki, S. Scharf, F. Faloona, K.B. Mullis, G.T. Horn, H.A. Erlich, and N. Arnheim, "Enzymatic amplification of beta-globin genomic sequences and restriction site analysis for diagnosis of sickle cell anemia", *Science,* vol. 230, no. 4732, pp. 1350-1354, 1985.
[http://dx.doi.org/10.1126/science.2999980] [PMID: 2999980]

[2]   N.W. Ohan, and J.J. Heikkila, "Reverse transcription-polymerase chain reaction: an overview of the technique and its applications", *Biotechnol. Adv.,* vol. 11, no. 1, pp. 13-29, 1993.
[http://dx.doi.org/10.1016/0734-9750(93)90408-F] [PMID: 14544807]

[3]   W.M. Freeman, S.J. Walker, and K.E. Vrana, "Quantitative RT-PCR: pitfalls and potential", *Biotechniques,* vol. 26, no. 1, pp. 112-122, 124-125, 1999.
[PMID: 9894600]

[4]   C.W. Dieffenbach, T.M.J. Lowe, and G.S. Dveksler, General Concepts for PCR Primer Design.*Genome Res.,* vol. 3, pp. S30-S37, 1993.
[PMID: 8118394]

[5]   P.D. Siebert, and J.W. Larrick, "Competitive PCR", *Nature,* vol. 359, no. 6395, pp. 557-558, 1992.
[http://dx.doi.org/10.1038/359557a0] [PMID: 1383831]

[6]   K.P. Foley, M.W. Leonard, and J.D. Engel, "Quantitation of RNA using the polymerase chain reaction", *Trends Genet.,* vol. 9, no. 11, pp. 380-385, 1993.
[http://dx.doi.org/10.1016/0168-9525(93)90137-7] [PMID: 7508648]

[7]   H. Mocharla, R. Mocharla, and M.E. Hodes, "Coupled reverse transcription-polymerase chain reaction (RT-PCR) as a sensitive and rapid method for isozyme genotyping", *Gene,* vol. 93, no. 2, pp. 271-275, 1990.
[http://dx.doi.org/10.1016/0378-1119(90)90235-J] [PMID: 1699848]

[8]   G.F. Gerard, "Inhibition of SuperScriptTM II reverse transcriptase by common laboratory chemicals", In: *Focus* vol. 16. Life Technologies, Inc.: Gaithersburg, MD , USA, 1994, no. 4, pp. 102-103.

[9]   P. Chomczynski, "A reagent for the single-step simultaneous isolation of RNA, DNA and proteins from cell and tissue samples", *Biotechniques,* vol. 15, no. 3, pp. 532-534, 536-537, 1993.
[PMID: 7692896]

[10]   D. Simms, P.E. Cizdziel, and P. Chomczynski, "TRIzol: A new reagent for optimal single-step isolation of RNA", *Focus,* vol. 15, no. 4, pp. 99-102, 1993.

[11]   A.M. Bracete, and D.K. Fox, "Isolation of total RNA from small quantities of tissues and cells", *Focus,* vol. 21, pp. 38-39, 1999.

[12]   A.M. Bracete, D.K. Fox, and D. Simms, "Isolation and Long-Term Storage of RNA from Ribonuclease-Rich Pancreas Tissue", *Focus,* vol. 20, no. 3, pp. 82-83, 1998.

[13]   C. Goerke, M.G. Bayer, and C. Wolz, "Quantification of bacterial transcripts during infection using competitive reverse transcription-PCR (RT-PCR) and LightCycler RT-PCR", *Clin. Diagn. Lab. Immunol.,* vol. 8, no. 2, pp. 279-282, 2001.
[PMID: 11238208]

[14]   H.A. Erlich, Ed., *PCR technology: principles and applications for DNA amplification.* IRL Press at Oxford Univ. Press: Oxford, UK, 1989.
[http://dx.doi.org/10.1007/978-1-349-20235-5]

[15]   S.R. Carding, D. Lu, and K. Bottomly, "A polymerase chain reaction assay for the detection and quantitation of cytokine gene expression in small numbers of cells", *J. Immunol. Methods,* vol. 151, no. 1-2, pp. 277-287, 1992.
[http://dx.doi.org/10.1016/0022-1759(92)90128-G] [PMID: 1629616]

[16]   M. Vaitilingom, F. Gendre, and P. Brignon, "Direct detection of viable bacteria, molds, and yeasts by reverse transcriptase PCR in contaminated milk samples after heat treatment", *Appl. Environ. Microbiol.,* vol. 64, no. 3, pp. 1157-1160, 1998.
[PMID: 9501455]

[17]   S.A. Bustin, V. Benes, J.A. Garson, J. Hellemans, J. Huggett, M. Kubista, R. Mueller, T. Nolan, M.W. Pfaffl, G.L. Shipley, J. Vandesompele, and C.T. Wittwer, "The MIQE guidelines: minimum information for publication of quantitative real-time PCR experiments", *Clin. Chem.,* vol. 55, no. 4, pp. 611-622, 2009.
[http://dx.doi.org/10.1373/clinchem.2008.112797] [PMID: 19246619]

[18]   Z. Wang, M. Gerstein, and M. Snyder, "RNA-Seq: a revolutionary tool for transcriptomics", *Nat. Rev. Genet.,* vol. 10, no. 1, pp. 57-63, 2009.
[http://dx.doi.org/10.1038/nrg2484] [PMID: 19015660]

SUBJECT INDEX

www.ingramcontent.com/pod-product-compliance
Lightning Source LLC
Chambersburg PA
CBHW041716210326
41598CB00007B/669